AF526224

PHASE AND CASTE DETERMINATION IN INSECTS

Endocrine Aspects

PHASE AND CASTE DETERMINATION IN INSECTS

Endocrine Aspects

Papers presented at a Symposium of the Section Physiology and Biochemistry of the XV International Congress of Entomology, Washington D.C., 1976 (Chairman: Dorothy Feir)

Edited by

MARTIN LÜSCHER

Universität Bern, Switzerland

PERGAMON PRESS

OXFORD · NEW YORK · TORONTO · SYDNEY · PARIS · FRANKFURT

U.K.	Pergamon Press Ltd., Headington Hill Hall, Oxford OX3 0BW, England
U.S.A.	Pergamon Press Inc., Maxwell House, Fairview Park, Elmsford, New York 10523, U.S.A.
CANADA	Pergamon of Canada Ltd., P.O. Box 9600, Don Mills M3C 2T9, Ontario, Canada
AUSTRALIA	Pergamon Press (Aust.) Pty. Ltd., 19a Boundary Street, Rushcutters Bay, N.S.W. 2011, Australia
FRANCE	Pergamon Press SARL, 24 rue des Ecoles, 75240 Paris, Cedex 05, France
WEST GERMANY	Pergamon Press GmbH, 6242 Kronberg/Taunus, Pferdstrasse 1, Frankfurt-am-Main, West Germany

First edition 1976

Library of Congress Catalog Card No. 76-20270

In order to make this volume available as economically and rapidly as possible the contributors' typescripts have been reproduced in their original forms. This method unfortunately has its typographical limitations but it is hoped that they in no way distract the reader.

Printed in Great Britain by A. Wheaton & Co., Exeter

ISBN 0 08 021256 5

Preface vii

Introduction. Martin Lüscher 1

Juvenile hormone and caste differentiation in the honey bee (Apis mellifera L.). Jan de Wilde 5

The role of determinator in caste formation in the honey bee. Heinz Rembold 21

Environmental, genetic and endocrine influences in stingless bee caste determination. H.H.W. Velthuis 35

Juvenile hormone and queen rearing in bumblebees. Peter-Frank Röseler 55

Endocrine control over caste differentiation in a Myrmicine ant. M. V. Brian 63

The influence of juvenile hormone analogues on caste development in termites. Ivan Hrdý 71

The dependence of hormone effects in termite caste determination on external factors. Michael Lenz 73

Evidence for an endocrine control of caste determination in higher termites. Martin Lüscher 91

Juvenile hormone and aphid polymorphism. Dinah F. Hales 105

Neurosecretory control of polymorphism in aphids. Colin G.H. Steel 117

PREFACE

The organization of a symposium on phase and caste differentiation goes back to a suggestion of the Committee of the section Physiology and Biochemistry (Chairman Dr. Dorothy Feir) of the International Congress of Entomology. It was first meant to cover the whole field of polymorphism in insects, but was restricted to the endocrine aspects in order to allow a deeper insight in one of the more important recent developments.

Most authors agreed that a publication before the conference would be advantageous. The only possibility of doing this, was by using the offset lithography procedure. Consequently the manuscripts had to be prepared camera-ready by the authors and some duplications could not be avoided. On the other hand, each article is complete in itself and provides all the background necessary for its understanding.

Two authors, Dr. D. Hales and Dr. M. V. Brian, had to cancel their participation at the Symposium. They agreed, however, to submit their papers for publication. Dr. G. B. Staal*, who for reasons beyond our control, was asked to report at the symposium at a very late stage, could not submit a manuscript.

My sincere thanks are due to Dr. Dorothy Feir for her help and advise in organizing the Symposium. I am grateful to Pergamon Press, who agreed to publish the Symposium in an incredibly short time.

Bern Martin Lüscher

* G. B. Staal: Juvenile hormone and phase development in locusts.

Errata

Introduction page 1, line 16 ''polyphenism'' should read ''polyethism''

Introduction page 1, line 23 ''lucusts'' should read ''locusts''

Rembold figure 2, page 24 ''25%'' should read ''25° C'' (twice)

INTRODUCTION

Martin Lüscher
University of Bern, Zoological Institute,
Division of Animal Physiology,
Engehaldenstrasse 6, CH-3012 Bern, Switzerland

Polymorphism or the ability of certain animals, particularly insects, to appear in more than one morphologically distinct forms or 'morphs' has always aroused much interest as a phenomenon of differentiation on a higher level. One major event in the history of insect polymorphism was a Symposium of the Royal Entomological Society of London, convened by J.S. Kennedy in 1961 (1). At this occasion there was some discussion about the definition of polymorphism and it was felt that the term should be reserved for cases with two or more distinct forms and should not include continuous variation. Clearly this definition was based on morphological aspects which were at that time the subject of most research. With the years more and more evidence has accumulated showing that the different morphs are also distinct in their physiology and that morphologically identical individuals can be physiologically very different or show behavioural differences. For the latter case the term 'polyphenism' has been widely used, but it seems to me that we need a term which includes all these related phenomena. The term 'polyphenism' (from Greek phainein, to appear) is herewith proposed and will be used in this introduction.

The distinct forms can appear all at the same time as in the castes of social insects, or they can appear in successive generations as in the phases of lucusts and in the morphs of aphids. The same individual can also change its morphology and/or its physiology drastically during its own life span. If we accept these changes as belonging to the same category of phenomena, then metamorphosis becomes a case of polymorphism, as postulated by Wigglesworth (2) or of polyphenism. The term polyphenism should also include the diversity of hive bees and field bees and of summer and winter bees. As will be seen, these phenomena are all controled by hormones and are thus closely related in their determination mechanisms.

It is perhaps typical for the time of the 1961 Symposium that the morphology and the external factors determining the development of morphs like nutritional factors, photoperiod, crowding (group effect), pheromones etc. were in the focus of the discussions. Only marginally there were some suggestions that these factors might influence the endocrine system which ultimately could bring about the specific developmental changes leading to different forms. Nothing was known then in that respect about the most important group of the social Hymenoptera. Even in the recent book by Schmidt on social polymorphism (3), which appeared 1974 but was completed as a manuscript in 1972, there is hardly a mention of hormone actions in caste development of Hymenoptera.

However, during the last 10 or 15 years great advances have been achieved in most other polyphenic insects and during the last 4 years also in the Hymenoptera. It has now become clear that in almost all carefully investigated cases the external factors influence the endocrine system and that hormones are ultimately responsible for determination. The rapid advances in this field were possible thanks to the availability of insect hormones and hormone analogues as well as to the now achieved high sensitivity of hormone assays.

As far as we know, the evidence for all but one group of insects (aphids) points to a crucial role of juvenile hormone (JH) in the determination of phases and castes. It is known for some time already that JH induces the characters of the solitary phase in locusts (4). In termites the role of JH in determining presoldier development was discovered in 1958 (5), but many years elapsed before the effects of JH on other caste developments could be elucidated (6).

In the social Hymenoptera the first conclusive experiments showing an influence of JH on queen development in the honey bee after application to the young larvae were reported in 1972 (7). Then it was soon discovered that the stingless bees were susceptible to become queens when treated with JH during the praepupal stage (8). In this case the determining corpora allata-activity is induced by the presence of certain genes. Similar effects of JH could be shown in larvae of bumblebees (9) and of one species of ants (10).

In bumblebees young queens treated with JH change their physiology and assume workerlike behaviour (9). This is one example of polyethism being under JH control. Another example is the change of the hive bee to the field bee. It was shown in our laboratory that this change is accompanied by a drastic increase of the JH titre in the haemolymph (Table 1) and that injected JH brings about physiological changes which are normally associated with the change to field bee behaviour like the breakdown of the pharyngeal gland and the decrease of the haemolymph protein concentration (11).

The change from the hive bee physiology to the physiological state of the field bee thus seems to be brought about by JH. In contrast to this, the change from summer bee physiology to that of the winter bee with an increase of the haemolymph protein titre is correlated with a drastic decrease of the JH activity in the haemolymph (12). It is therefore possible that the physiological state of the winterbee is brought about like diapause in some other insects by an inactivation of the corpora allata.

TABLE 1 JH-Titres in Haemolymph of adult Worker Honey Bees, after Rutz (11) and Fluri (12)

	Age	Galleria-Units per ml
Summer bees May - August	12-24 hours	328 ± 42
	12 days	1807 ± 249
	24 days	3820 ± 360
Winter bees November - January	40-130 days	< 250 (not measurable)

In all these polyphenism cases JH plays the most important role in causing determination. In aphids there is some evidence that JH is involved in the determination of apterous morphs (13) but experimental results are not quite conclusive. However, in the case of <u>Megoura</u> the production of sexually or asexually reproducing progeny is determined in the mother by daylength which influences neurosecretory cells. These release their secretory product which is then transported intra-axonally to the reproductive system (14). The progeny are then determined by this secretion to become virginoparae. This seems to be a special case amongst the polyphenic insects, but it may well be that neurosecretory mechanisms will also be discovered in other cases.

We have now seen that the external determining factors act on the endocrine system which then induces the final determination. There is, however, one external factor, the role of which may be different, the so-called queen bee determinator (15) isolated from royal jelly. It cannot be a nutritional factor since it is active in incredibly small amounts. It may itself be an endocrine principle, since it can be extracted from other insects where it may be involved in some step of normal development. It may be a substance which triggers and activates the corpora allata. We have therefore included it in our discussions in the hope that the situation will be clarified.

Hormone research in relation to polyphenism is still at its beginning. Let us hope that this symposium will stimulate further research in this fascinating field.

REFERENCES

(1) J. S. Kennedy, Ed. Insect Polymorphism. Symp. Roy. Entomol. Soc. London, 1, 1-115 (1961).

(2) V. B. Wigglesworth, Insect polymorphism - a tentative synthesis. In (1) 103-113.

(3) G. H. Schmidt, Sozialpolymorphismus bei Insekten - Probleme der Kastenbildung im Tierreich. Wiss. Verlagsges. Stuttgart 1-974 (1974).

(4) G. B. Staal, Studies on the physiology of phase induction in Locusta migratoria migratorioides. Thesis, H. Veenman and Zonen, Wageningen 1-124 (1961).

(5) M. Lüscher, Experimentelle Erzeugung von Soldaten bei der Termite Kalotermes flavicollis (Fabr.), Naturwiss. 45, 69-70 (1958).

(6) K. Wanyonyi, The influence of the juvenile hormone analogue ZR 512 (Zoecon) on caste development in Zootermopsis nevadensis (Hagen) (Isoptera), Insectes Sociaux 21 35-44 (1974).

(7) P. Wirtz and J. Beetsma, Induction of caste differentiation in the honey bee (Apis mellifera L.) by juvenile hormone. Ent. exp. et appl. 15, 517-520 (1972).

(8) H.H.W. Velthuis, this volume, 35-53.

(9) P.-F. Röseler, this volume, 55-61.

(10) M. V. Brian, this volume, 63-70.

(11) W. Rutz, L. Gerig, H. Wille and M. Lüscher, On the function of juvenile hormone in adult worker honeybees (Apis mellifera), J. Insect Physiol., in press.

(12) P. Fluri, H. Wille, L. Gerig and M. Lüscher, Juvenile hormone and protein titres in haemolymph of winter-honeybees, in preparation.

(13) D. F. Hales, this volume, 105-115.

(14) C.G.H. Steel, this volume, 117-130.

(15) H. Rembold, this volume, 21-34.

JUVENILE HORMONE AND CASTE DIFFERENTIATION IN THE HONEY BEE (APIS MELLIFERA L.)

Jan de Wilde
Department of Entomology, Agricultural University, Wageningen
The Netherlands

1. INTRODUCTION

When Boerhaave published the collected works of Jan Swammerdam in *Biblia Naturae* (1) it was found that this remarkable naturalist had for the first time given a scientific description of the true nature of the Honey Bee queen. He had even observed the resemblance between queen and worker bee. Of the last category he states "...that the common Worker Bees have much more in common with the character and nature of the Female than with those of the Male". Moreover, in 1669, Swammerdam had expressed clear and original views on the process of insect metamorphosis. In *Historia Insectorum Generalis* (2) he had laid down the principle of what he named Preformation and Epigenesis. In describing how in the Cabbage White, the wings develop from "anlagen" already present in the larva, Jan Swammerdam refers to the previous statements by William Harvey (3) who in his *DeGeneratione Animalium* had expressed the view that adult organs emerge "de novo" from the pupa, which he considered to be "a soft egg". In courteously refuting Harvey's argument, Jan Swammerdam states ..." that insects are difficult to study, unless by persons accustomed to experiments after this kind; it is no wonder that the most happy geniuses, the immortal Harvey, for example, and many others, should have fallen into error". It now appears that in the study of caste formation in the Honey Bee, we find recent examples of wrong conceptions, emerging from unawareness of the complexity of ontogenetic determination; but this time the misunderstanding involves the causal mechanisms rather than development itself.

The first important advances in our knowledge on Honey Bee castes after Swammerdam are due to Dzierzon (4) and von Berlepsch (5). These authors proved beyond doubt that both queen and worker emerge from the same type of egg, the fertilized egg, prone to develop into an individual of the female sex. And after Nachtsheim (6) gave conclusive cytological evidence, it was realized that drones, workers and queens carry the same quality of genetic information, but that drones develop from eggs carrying merely half the quantity as compared with those from which the female sex emerges. Several theories have been launched to explain how sexual differentiation can be based on merely quantitative differences; this aspect will not concern us here. In this paper, we will discuss our present views on how gene expression in the female Honey Bee larva is directed towards either of its two oecomorphs [*)], the queen or the worker.

* The term "morph" was introduced by Hille Ris Lambers (7) in a review on polymorphism in Aphids. For reasons of clarity, I have used the term *oecomorph* (8) to denote polymorphism in genetically identical individuals, as induced by extrinsic conditions.

2. THE DETERMINATION OF CHARACTERS AT METAMORPHOSIS

By experiments such as those of Geigy (9) it is known that ontogenetic determination of adult characters in insects takes place during early cleavage stages, when the blastoderm is being formed. The determination of cephalo-caudal polarity and left-right symmetry most probably takes place already during oogenesis. Determination of embryonic structures is effectuated by a "differentiation centre", inducing certain parts of the blastoderm to develop into an organized "germ band", in which the blue print for the future larva is present. In Holometabola, the larva carries a system of imaginal discs from which parts of the adult epidermal structures (and in Hymenoptera virtually major part of the adult integument) is reconstructed during metamorphosis. Adult differentiation in Holometabola, therefore, takes the shape of a program of programs.

It is not known, which factors determine the prospective significance of an imaginal disc, its development into the corresponding organ or appendage. Mention is made in literature of "position effects" (10). The development of cuticular patterns by the epidermis, so important in differentiating between queen and worker bee, is thought to be governed by gradient effects, in which communication between the epidermal cells is extensive (11). The occurrence of *transdetermination* when an imaginal disc is isolated and cultivated during a considerable time (14) and the fact that the cells of such a disc, when separated and recombined, reorganize their "normal" order and produce well-coordinated parts of organs and structures (13) proves that the program is carried out by the interaction between intracellular and extracellular factors. Such "determination factors" are present in the cortical plasm of the egg and determining messages are later exchanged between the ectodermal cells; they never provene from the environment.
In an insect such as the Honey Bee, morphogenesis implies the differentiation of at least three categories of characters:

1. Characters determining the topographic differentiation of the body: head-thorax-abdomen.
2. Characters specific to the caste.
3. Characters specific to the developmental stage or instar.

Although the developmental program sets prohibitive limits to the shape and size of characters at metamorphosis, some freedom is left for interference by the internal milieu, and sometimes by the external environment. This is already apparent by the polymorphism demonstrated by *stages* and *instars* of postembryonic development; it is, moreover, shown by the occurrence of "oecomorphs" such as *phases* in locusts, *castes* in social insects, and aphid *morphs*. The syndromes constituted by these types of polymorphism bear a clear relation to specific ecological requirements. The factors involved here, do not "determine" development; they provide tokens or messages, which, after being led through several steps of translation, modify gene expression. Among the most prevailing environmental tokens are those found in photoperiod, temperature and food; but the developmental program determines the pattern of genes which may ultimately be turned on. The use of the term "Determinator" to denote a substance in Royal Jelly which is thought to provide messages effective in queen formation (14, 15) is therefore ill advised.

3. ENDOCRINE IMPACT ON INSECT DEVELOPMENT

After Kopec (16) found that the brain of an insect larva displays incretory activity initiating moulting and metamorphosis, Wigglesworth (17, 18) concluded to the existence of a "Juvenile Hormone" produced by the Corpora allata and Fukuda (19) and Williams (20) elucidated the rôle of the prothoracic glands, the picture of the endocrine regulation of moulting and metamorphosis was further enriched by E & B Scharrer (21) who introduced the concept of neuro-endocrine integration.
For a considerable time, it seemed that the endocrine control of morphogenesis was fully explained by a recurrent wave of activity of the medial neurosecretory cells of the brain, followed by activation of the prothoracic glands, and (only in case of a larval or pupal moult) of the corpora allata. The subsequent isolation and identification of Ecdysone (22, 23) and of the Cecropia I Juvenile Hormone (24) merely seemed to confirm this state of events.

While the rôle of Ecdysone has not been disputed on the basis of later evidence, and was merely extended (25) (although the question of its site of production is more and more a matter of discussion, 26, 27, 28), our ideas on the rôle of the Juvenile Hormone have been subject to both extension and modification.
As regards the "Juvenilizing effect", Williams (29) merely conceived an inhibitory "status quo" function, Wigglesworth (30) an active induction of larval or pupal characters, while Slama (31) and some other authors in recent literature emphasized its antigonistic effect versus Ecdysone, either its production or its activity. It was shown that an increase in JH titre may delay a moult, and may even lead to larval diapause (32). The course of the JH titre during the period preceding metamorphosis in caterpillars suggests a very intricate interplay, coinciding with behavioural phases in preparing pupation sites (33). It was also shown that JH plays a role in the induction of phase characters in Locusts (34) and of caste characters in termites (35). By all these discoveries, the term "Juvenile Hormone" became more and more questionable. Juvenile Hormone now appears to be a versatile hormone, used by insects in regulating the timing of moulting, the state of morphogenetic differentiation, elements of behaviour (36), the synthesis or atrophy of flight muscles (37, 38, 39) and several aspects of the reproductive cycle. Several of the above activities profoundly involve the rate of metabolic activity. Very important new data were provided by Naisse (65) who demonstrated that in the Glowworm, *Lampyris noctiluca*, sex determination involves the action of both corpora allata and corpora cardiaca, and can be modified by interfering with these glands.

The availability of pure JH preparations and of a large number of JH bioanalogues, enabled further progress. It was found that in the Insect Class at least three different JH's occur, named after the sequence of their discovery JH I, II and III. The last-named hormone, JH III is according to Trautmann et al. (41) the JH of the adult Honey Bee.
Very interesting data were obtained by JH application at different moments during the life cycle. It was proved possible to interfere with the normal developmental program in such a way that the effects became apparent a considerable time after application. Such "delayed" effects occurred, e.g. after treatment of the adult female (disturbance of embryogenesis in the

eggs produced) and upon treatment of the developing embryo (disturbance of larval and pupal development) (42).

Some of the above data led us, in 1969, to investigate whether in the Honey Bee, activation of the corpus allatum and a corresponding increase in JH titre early in larval life might lead to "covert" morphogenetic induction, directing the developmental program towards queen differentiation, and becoming "overt" later in development. Some aspects of the rôle of the corpora allata in the regulation of larval-pupal differentiation had been investigated experimentally by Lukoschus (43). Suggestions for a differential activity of the corpora allata in queen and worker were derived in experiments of Wang & Shuel (44) and Dixon & Moser (45).
The inhibition of adult differentiation by JH analogues, applied to the later developmental stages of the Honey Bee was observed by Zdárek & Haragsim (46) who also confirmed the ind ction of queen-like characters by such treatments.

4. HONEY BEE CASTES AS OECOMORPHS

In the Honey Bee, the female sex is expressed in two phenotypes which differ in a number of important respects: Morphological, anatomical, physiological, and behavioural.
Morphological differences concern absolute and relative size as well as shape of body parts: head, abdomen, appendages, and the detailed structure and chaetotaxy of appendages, especially the hind legs.
Anatomical differences include the brain (hypertrophy of associative centres in the worker's protocerebrum) and the gonads. In the worker,the spermatheca is rudimentary and the number of ovarioles amounts about 2% of the number found in the queen. There are also marked differences in the development of glandular structures, such as include activity of the pharyngeal glands, as well as wax glands and Nassanoff glands which are present in the worker, but absent in the queen. The worker's mandibular glands secrete components of larval food, while in the queen, the mandibular glands secrete a.o. a pheromone suppressing the workers' reproductive activity, at the same time being a sex-attractant to drones. Endocrinologically it has been reported that the activity of the corpus allatum is larger in the adult worker than in the queen, as judged by biometrical data (47, 48). Lukoschus (47) gives a more or less extensive survey of the different caste characters.

While in the normal colony the queen displays a high rate of oogenesis, this is not observed in the worker's ovarioles.
Behaviourally, there is a wealth of differences; the complicated set of social behaviour elements in the worker bee, including a "language"; its long-lasting memory and its capacities of conditioning, as compared with a much reduced pattern in the queen, predominantly amounting to reproductive behaviour.
Especially the remarkable division of labour between the castes and among the worker bees, the last-named combining most unusual foraging architectural and storage capacities, have made the honey bee an outstanding example of caste formation.
The two castes differ, according to the above, in two main respects:

1. Differences emerging from differential development.
2. Differences emerging from differential functioning of existing organs depending on the position of the two castes within the structure of the colony.

As the two castes represent phenotypes based upon one and the same genome and are induced by differential treatment during larval development -the nutritive environment of the larva being the main factor- the term *oecomorph* applies. For practical reasons, it is important to make use of discriminating characters in experiments where, by differential manipulation in the colony or in vitro, the caste characters are not always expressed in their extreme form, but show more or less continuous transitions. Such discriminating features are: notching of the mandibles, chaetotaxy of tibia and basitarsus, number of ovarioles, size of the spermatheca, ratio of lateral and medial venom duct, presence of hypopharyngal glands. From recent experiments (50) however, we know that the rate of queenlikeness of induced queens never can be determined on the basis of a few discriminating criteria, but that it is necessary to take into account a combination of different characters. Inclusion of behavioural criteria such as retinue and aggressive behaviour among induced queens renders the discrimination even more conclusive.

When it now comes to the question, which of the two castes in the Honey Bee deviates most from the normal female development pattern in solitary *Apidae*, I can only agree with Lukoschus (48) that this is certainly the worker. The development of a mass-rearing system for female worker individuals with a reduced reproductive system which is extrinsically inhibited by the presence of the queen, and the above-mentioned exuberant diversity in behaviour belonging to the potential of these workers, is the real novelty in the evolution of social *Apidae*. Rather than concentrating on factors promoting reproductives, our problem therefore seems to be: How is the female larva manipulated to become a worker?

5. ONTOGENESIS IN THE TWO CASTES

The timing of moults in the two castes, according to Bertholf (51) is given in Fig. 1.

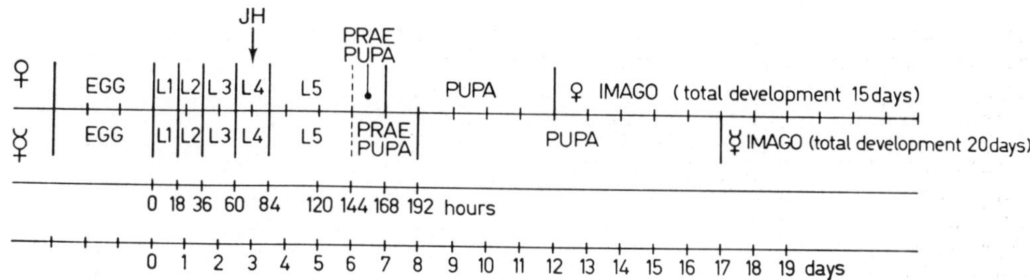

Fig. 1. Time-table of postembryonic development of queen and worker larva (after Berthold, 1925). The arrow indicates the time at which the developmental program can be switched from worker into queen development by JH application.

Moulting occurs synchronously until L5; thereafter a difference ensues. While spinning starts about the 6th day of larval life in both cases, the pupal moult of the queen occurs one day earlier than in the worker. The difference of one daily cycle is probably real; the "gate" for the pupal moult is most

likely a circadian event. The most important difference in development timing is in the duration of the pupal stage; 5 days in the queen, 9 days in the worker. The regression of developmental rate in the worker caste, which apparently sets in before the beginning of cocoon spinning is most dramatic in the ovaries. While at 72 hrs, the full set of ovariole primordia is present in both castes and cyst formation is starting, ovarian regression begins in the worker larva, reducing the number of germarial "anlagen" in each ovary. At 96 hrs some 4-6 primordia are left, which subsequently develop into germaria. The queen retains her full number of 150-180 germarial primordia per ovary. These regressions in worker development are the first outward signs of a program which must be defined in the early larval stages, but here the differences between the future castes are mainly covert; we mostly observe quantitative differences in growth rate, nutritive state, and corpus allatum activity.
Recently Goewie (52) reported a morphological difference in labial chemoreceptors in presumptive larval queens and workers. From 72 hours on, when the size of larval receptors increases, queen larvae develop labial sensilla styloconica with 3 conical papillae. Worker larvae on the contrary merely develop 2 conical papillae. (Fig. 2).

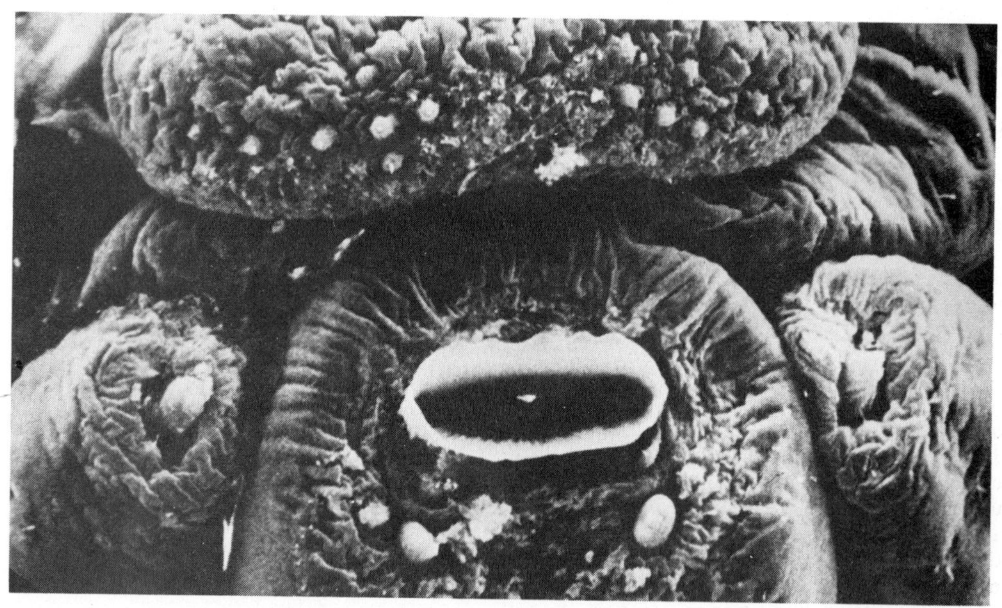

Fig. 2. Scanning electronmicroscopical picture of the mouthparts of the honeybee larva. The labium with the two sensilla styliconica situated ventro-laterally of the common opening of the silk glands is seen below. After Goewie (52)

In fact this is the earliest outward sign of caste differentiation in Honey Bee larvae. The possible significance of this structural difference in relation to the fixation of caste differentiation is under study.
In the following pages we will discuss why the endocrine activity of the corpora allata constitutes a decisive factor in switching the program, and how during development, different elements of morphogenesis are successively brought under the control of these glands: first, the features of caste; later, the features of the ultimate developmental stages.

The nutritive state of the larva during the early phase of development deserves special mentioning. Fat body cells of worker larvae during the first three days, according to the histological picture, contain little stores and have large watery vacuoles. Only after 3½ days, during the "gorging phase" (51), when modified worker jelly is being fed, accumulation of glycogen and fat occurs. Larvae in queen cells, on the contrary, right from the beginning accumulate abundant stores of glycogen in their fat body (53). Wirtz concludes that worker larvae are "left short of food" during the first three days of their life. This observation does not explain, however, why during the first three days the weight of the worker larva does not differ significantly from that of a presumptive queen larva.
The synchronism between the moults of presumptive queen and worker larvae indicates that nutrition in both categories is sufficient to pass the "periode de nutrition indispensable" (54) which is essential for a moult. It follows that nutrition of early worker larvae is delicately balanced.

6. CASTE INDUCTION AND THE "DETERMINATOR" PRINCIPLE

In the Honey Bee, larval nutrition during the first three days determines in which direction the developmental program will be switched: towards the queen or the worker. This fact has led several scientists to investigate which component in the diet is responsible for the switch. Both Rembold (15, 55) and Weaver (56, 57) reported that, when Royal Jelly is fractioned into several components, the dialysable fraction contains a substance which is indispensable for larvae reared in vitro on the recombined diet, to develop features of a queen.
From literature (55,59) one obtains the impression that this substance was originally thought to be directly involved in the morphogenetic process, not unlike vitamins are involved in metabolic processes. Hence the term Determinator which was attached to the active principle.
In view of the very specific differences between queen and worker bees, it was conceived that the action of the substance would also be highly specific. And when, in a later phase (14,55) dialysable fractions prepared from silkworm pupae, when added to a worker jelly diet, contributed to switch the development of a female larva reared <u>in vitro</u> in the direction of queen formation, it was conceived that the Determinator was also present in the silkworm, and therefore was a differentiating principle of more general occurrence among insects.
Quite apart from the question whether or not it is likely that the differentiation of reproductives needs a specific stimulus (p.7), ontogenetic determination in general is not directly subordinated to environmental factors.
Moreover, after the findings of Wirtz (53) it is most likely that Royal Jelly contains one or more substances triggering the activity of the corpora allata, probably in an indirect manner. It is likely that the above mentioned dialysable substance takes its place in this category, and does not represent a separate morphogenetic principle.

7. CASTE PROGRAMMING AND THE RÔLE OF JUVENILE HORMONE

Until 1973 it was known that the corpora allata in presumptive queen larvae are considerably larger than those of presumptive worker larvae (62). But in relation to body weight the difference appeared to be less conspicuous (63) or even nonexisting (64). It was Wirtz (53), Wirtz and Beetsma (60) who, in a series of conclusive experiments, proved beyond doubt that caste programming is under the control of the corpus allatum. The evidence which these authors accumulated can be summed up as follows:

1. first, it was shown, upon grafting worker larvae into queen cells, that the optimum age, at which the volume of the corpus allatum of these larvae increases fastest in response to the new conditions, is at the end of the 3rd day (Fig. 3).

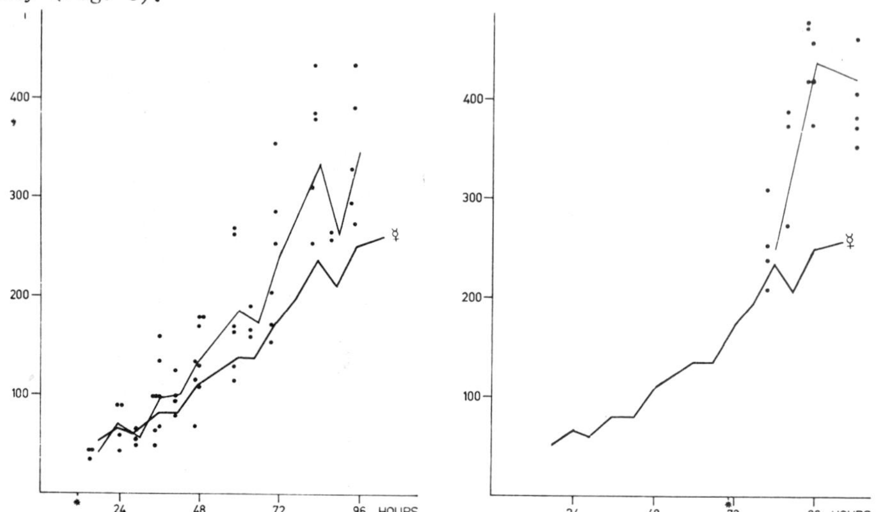

Fig. 3. Surfaces of corpora allata of queen larvae grafted at the age of
left: 5 to 18 hours (first instar)
right: 70 hours (fourth instar)
as compared with worker larvae. * moment of grafting

2. Subsequently it was shown that the ultrastructural activity pattern in corpora allata of presumptive queen larvae shows characteristic differences from that in presumptive worker larvae, pointing towards a much higher specific rate of synthesis of JH in glands of presumptive queen larvae. The difference diminishes at the age of 123 hours and has disappeared at cocoon spinning, at the beginning of the prepupal phase.

3. It was subsequently shown that the JH titre in the haemolymph of presumptive queen larvae is much increased as compared with worker larvae. When worker larvae 44 hrs old were grafted into queen cells, at the age of 72 hrs the titer was more than ten times that in comparable worker larvae. It remained at about the same level until the moment of cocoon spinning, when it dropped to a level equal to that in the worker prepupa.

4. Wirtz finally resulted in mimicking the effect of the larva's own corpus allatum by external application of JH I to worker larvae. The optimal degree of expression of queen characters, which could be obtained under these conditions, was reached when the larvae were treated with 1 μg JH I shortly before the end of the third day. The worker bees continued to feed these larvae in the same way as worker brood, and did not discriminate between treated and untreated larvae until the time of cocoon spinning. After that time, they tended to remove the treated individuals, which therefore had to be incubated separately in a thermocabinet.

It follows from the above data that a high titre of JH switches the developmental program in such a way that the expression of caste characters is directed towards queen formation (Fig. 4).

Fig. 4. Effect of juvenile hormone application on development of the mandibular tooth
left: swarm-queens, middle: JH induced queens, right: worker.

This includes the shortening of development which is characteristic for the queen; and this is one of the most convincing arguments in favour of the program concept. For whatever we know of the immediate effect of JH on the timing of a moult points to a delay rather than acceleration.

As in the experiments by Wirtz, the larvae were reared in worker cells, they were small; also intercaste features were abundant. Queen expression was, therefore, still imperfect. This was further investigated by Goewie (50) who, as a sequel to the findings of Trautmann et al. (41) repeated the experiments with JH III. As is shown in Fig. 5, a much more optimal queen expression is induced by this hormone especially after repeated topical application. Very interesting data on the program of ovarian development were obtained by Naisse(unp.).She could demonstrate that JH I can still prevent the atrophy of ovariole primordia in worker larvae until the 4th day of development; after that time, a progressively decreasing number of primordia is still left for development. After day 4, no further change is possible. True *reversion* of the program was only observed in some cases where cysts already showed a developmental arrest, and were activated by the hormone. But cysts were never observed to reappear where they had already partly or wholly atrophied.

Some of the above morphogenetic effects induced by pure JH were obtained by other workers (46) and in our laboratory (66) with JH bioanalogues. It is worthwhile mentioning here that the *functioning* of castes is also under the influence of JH. This is especially interesting in the worker bee, where the functioning of the hypopharyngeal glands (67) and of the wax glands is greatly reduced by high JH titres.

Finally, it is important to reflect on the *primum movens* of the internal

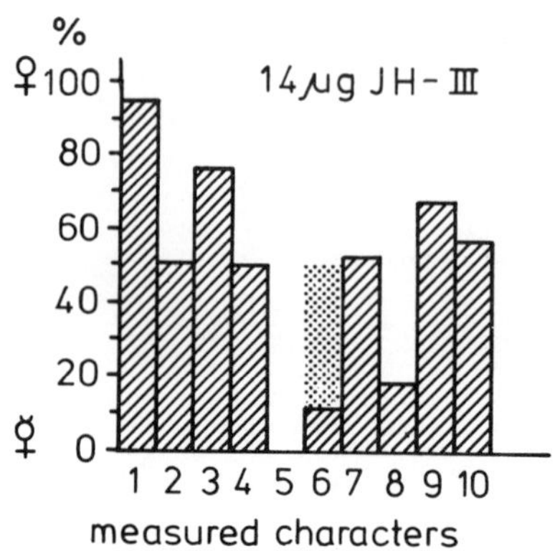

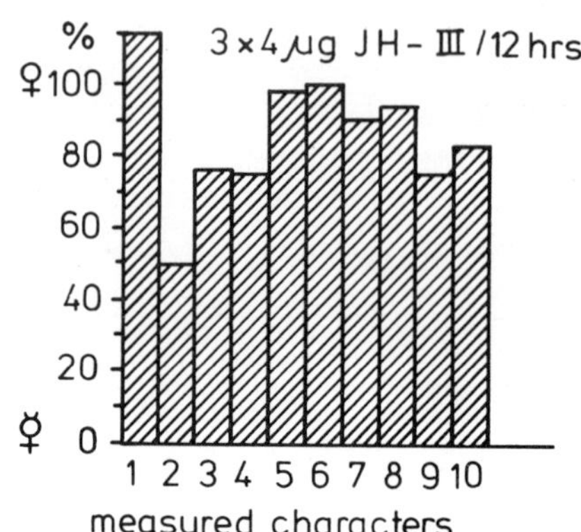

Fig. 5. Effect of repeated JH-III applications.
The value of each parameter has been expressed as a percentage of the value found in swarm queens. After Goewie (50)

1. length of glossa
2. development of mandibular tooth (notching)
3. ratio of width and length of the head
4. size of mandibular gland
5. chaetotaxy of tibia
6. chaetotaxy of basitarsus (dotted area indicate number of incomplete hair rows)
7. number of barbs on the lancet
8. number of ovarioles per ovary
9. size of spermatheca
10. ratio of median and lateral venom gland duct.

events in caste induction. Certainly this is not the corpus allatum, as this is controlled by neuro-endocrine integration, the centre of which is located in the brain. Neurosecretory activity has been studied ultrastructurally to some extent by Wirtz, and is certainly present already in the early larval stages. But quantitative measurements are still wanting.

8. IN SEARCH OF THE RELEVANT MESSAGE

When we now come to the question, which environmental trigger inhibits corpus allatum activity in presumptive worker larvae, and thereby reduces the juvenile hormone titre to a level merely allowing for the development of a worker bee, it seems indicated to take a completely open attitude. The most obvious factor would be limitation of food supply, as this is known to inhibit neurosecretory activity (and thereby, corpus allatum activity) in many insect species (68, 69, 70, 71, 72). And indeed, several observations point to the fact that during the first three days the worker larva is kept short of food. But a low rate of nutrition can be derived from qualitatively reduced food supply as well as from diminished "appetite", which again may be influenced by the composition of the diet and by the behaviour of the worker bees during feeding.

Jung-Hoffmann (73) found that worker larvae are fed variable amounts of white (mandibular) component. After 3½ to 4 days worker larvae store large amounts of food, and at the same time seldom receive the white component.
Shuel and Dixon (74) added sugar to worker jelly and thereby enabled worker larvae to enter the "gorging phase", which is a condition for worker larvae to develop into adults. They remark "One might speculate on a possible rôle of sugar in caste determination..... when considered in relation to the striking difference between the sugar concentration of early queen and worker diets".
These observations render it attractive to consider the possibility of a balance between a phagostimulatory component(invert sugar)and a slightly deterrent component (white mandibular gland secretion) in the diet. Several data in literature render it worthwhile to investigate this possibility.
Wirtz (53) has found that a sugar receptor is present on the buccal parts of the larva.
Lensky (75) describes that larvae reared in vitro on worker jelly with 16-20% of invert sugar added, developed into adult queens, intermediates, and workers.
In the above context, it seems urgent to study the sensory effect of mandibular gland secretion on larvae, and, moreover, the effect on their feeding behaviour. In addition to the above considerations, we may be mindful to the conclusion derived by Jung-Hoffmann (73) that it is not only important *when* certain food components are offered, but also *how* they are offered by the worker bees.

9. DISCUSSION

In this discussion I shall restrict myself to the literature pertaining to the Honey Bee, and will not discuss caste induction in other social *Hymenoptera*, although it would seem that similar conditions such as we have recently detected, also prevail in Bumblebees (76) and *Meliponidae* (77).
Working with Honey Bees is, as we know, very much "a way of life".
In retrospect, it would seem that work on caste determination in the Honey Bee has thereby suffered from taking an isolated position in the field of insect developmental physiology, which in recent years has undergone such quick advances. Statements like "The corpus allatum is the only well-developed endocrine organ in the early larval stages" (44) and the conception that neuroblasts in the brain of Honey Bee larvae have a neurosecretory function (62) stand alone in insect endocrinology, a fact not sufficiently realized by the scientists quoted.
The same deficiency is observed in many of Rembold's papers, e.g. where the influence of JH is explained as an indirect morphogenetic effect not directly coupled to caste induction (80) or where he denies the effect of JH in queen formation, because Dixon and Moser (45) observed that at 48 hrs, implantation of corpora allata of queen larvae had a repressive effect on corpus allatum activity in roaches. The findings of Dixon and Moser (45) are easily explained by the fact that corpus allatum activity is subject to feedback regulation.

Many interesting experiments on the functioning of endocrines in Honey Bee larvae have been carried out, and I would by no means doubt the exactitude of painstaking experiments and observations such as those by Wang and Shuel (44), Ritchey and Dixon (78), Dixon and Moser (45). My objections merely pertain to the deficiencies in interpretation signalled above.
Among the endocrinologically acceptable discussions on the above theme, those originating from the school of Piepho have been too infrequently

quoted in literature on caste induction. I am referring to the papers by Lukoschus. In several of his contributions (47, 48, 63, 79) this author not only places the developmental events in the right framework, but some of his considerations have brought him very near to discovering the mechanism which is now gradually being unraveled in our laboratory. His prophetic paper (48) emphasizes the crucial points and could well have prevented other workers from searching inadvertantly for "Determinators" of caste development.

Although workers such as Rembold et al. (80) seem to have great difficulty to agree with the impact of the early JH effect in caste induction in the Honey Bee, they do not object to the findings of Lüscher (81) and Wanyonyi (82) that JH in termites induces the differentiation of larvae into presoldiers. It would seem to us that the reason for this difference in appreciation is the short-term effect of JH in caste induction in termites as compared with the apparently long-term effect in queen induction in the Honey Bee.

In termites, presoldier induction is dependant on an increased JH titre during the intermoult period of the larva, after which the presoldier is formed; in the Honey Bee on the contrary, a queen emerges after some 9 days, and after three moults have elapsed.

But here we should be mindful of the difference between the hemimetabolous termites, which from a point of view of developmental physiology are close to the *Blattidae*, and the holometabolous *Hymenoptera* with a metamorphosis based on imaginal discs, and a correspondingly complex developmental program. Only this complexity has made it possible first to let the corpus allatum preside over the characters of caste, and only at the final phase of development over the expression of pupal or adult characters.

ACKNOWLEDGEMENT

It is my pleasure to thank Mr. J.Beetsma and Mr. E.A.Goewie for useful remarks.

REFERENCES

(1) J.Swammerdam, Biblia Natural, Amsterdam (1737/1738).
(2) J.Swammerdam, Historia Insectorum Generalis, Utrecht (1669).
(3) W.Harvey, Excercitatones de Generatione Animalium (1651).
(4) J.Dzierzon. Papers on fertilization and parthenogenesis in the Honey Bee. Eichstädt Bienen-Zeitg. 1-32 (1845).
(5) A.von Berlepsch, Die Biene, und ihre Zucht mit beweglichen Waben, 2nd ed. Mannheim (1869).
(6) H.Nachtsheim, Zytologische Studien, Arch.f.Zellforsch. 11, 169 (1913).
(7) D.Hille Ris Lambers, Polymorphism in Aphididae, Ann.Rev.Entomol. 11, 47 (1966).
(8) J.de Wilde, An endocrine view of metamorphosis, polymorphism and diapause in insects, Amer.Zool. 15, 13 (1975).
(9) R.Geigy, Erzeugung rein imaginaler Defekten durch ultraviolette Eibestrahlung bei Drosophila melanogaster, Wilh.Roux Arch.Entw.Mech.Org. 125, 406 (1931).
(10) W.J.Ouweneel, Determination, regulation and positional information in insect development, Acta Biotheoretica, XXI (1/2), 115 (1972).
(11) P.A.Laurense, Polarity and patterns in the postembryonic development of insects, Adv.Insect Physiol. 7, 197 (1970).
(12) E.Hadorn, Problems of determination and transdetermination, Brookhaven Symp. 18, 148 (1965).
(13) H.Tobler, Zellspezifische Determination und Beziehung zwischen Proliferation und trans Determination in Bein und Flügel Primordien von Drosophila melanogaster, J.Embryol.exp.Morphol. 16, 609, (1966).
(14) H.Rembold, Biochemie der Kastenbildung bei der Honigbiene, Naturwiss. Rundschau, 26 (3), 95 (1973).
(15) H.Rembold, B.Lackner ana J.Geistbeck, The chemical basıs of honeybee (Apis mellifera) caste formation. Partial purification of queen bee determinator from royal jelly, J.Insect physiol. 20, 307 (1974).
(16) V.B.Wigglesworth, Factors controlling moulting and metamorphosis in an insect, Nature, 135, 725 (1934).
(17) V.B.Wigglesworth, The functions of the corpus allatum in the growth and reproduction of Rhodnius prolixus (Hemiptera), Quart.J.micr.Sci. 79, 91 (1948).
(18) V.B.Wigglesworth, The action of growth hormones in insects, Symp.Soc. exp.Biol. 11, 204
(19) S.Fukuda, Induction of pupation in Silkworm by transplanting the prothoracic gland, Proc.imp.Acad.Japan. 16, 417 (1940).
(20) C.M.Williams, Physiology of insect diapause. II. Interaction between the pupal brain and prothoracic glands in the metamorphosis of the giant silkworm Platysamia cecropia, Biol.bull. 93 (2), 89 (1947).
(21) E.and B.Scharrer, Neuroendocrinology, Columbia Press, New York-London (1963).

(22) A.Butenandt and P.Karlson, Über die Isolierung eines Metamorphose-Hormons der Insekten des Insekten in kristallisierter Form, Z.Naturforsch. 9, 389 (1954).
(23) P.Karlson, H.Hofmeister, W.Hoppen & R.Huber, Zur Chemie des Ecdysons, Ann.Chem. 662, 1 (1963).
(24) H.Röller, K.H.Dahm, C.C.Sweely, B.M.Trost, The structure of the juvenile hormone, Angew.Chemie 6, 179 (1967).

(25) G.S.Fraenkel, Interactions between Ecdysone, Bursicon, and other endocrines during puparicum formation and adult emergence in flies. Am.Zool. 15, 29, (1975).
(26) F.Romer, Häutungshormone in den Oenocyten des Mehlkäfers, Naturwiss. 58, 324 (1971).
(27) J.I.Gilbert and D.S.King, Physiology of growth and development: Endocrine aspects, In: M.Rockstein, Physiology of Insecta, 1, 250, Academic Press, New York-London, 2nd ed. (1974).
(28) T.H.Hsiao, C.Hsiao and J.de Wilde, Moulting hormone production in the isolated larval abdomen of the Colorado beetle, Nature, 255 (5511), 727 (1975).
(29) C.M.Williams, The juvenile hormone. I. Endocrine activity of the corpora allata of the adult cecropia silkworm, Biol.Bull. 116 (2), 323 (1959).
(31) K.Slama, Some old concepts and new findings on hormonal control of insect morphogenesis, J.Insect Physiol. 21, 921 (1975).
(32) M.S.Fukaya and M.Kobayashi, Some inhibitory action of corpora allata in diapausing larvae of the rice stemborer, Chilo suppressalis, Appl.Ent.Zool. 1, 125 (1966).
(33) L.Varjas, P.Paguia and J.de Wilde, Juvenile hormone titres in penultimale and last instar larvae of Pieris brassicae and Barathra brassicae, in relation to the effect of juvenoid application. Experientia, 25, 213 (1969).
(34) G.B.Staal, Studies on the physiology of phase induction in Locusta migratoria migratoroides R and F. Med.Lab.Entomol.Landbouwhogeschool Wageningen, 72 (1961).
(35) M.Lüscher and A.Springhetti, Untersuchungen über die Bedeutung der corpora allata für die Differenzierung der Kasten bei der Termite Kalotermes flavicollis F. J.Ins.Physiol. 5, 190, (1960).
(36) J.W.Truman and L.M.Riddiford. Hormonal mechanisms underlying insect behaviour, Adv.Insect Physiol. 10, 297 (1974).
(37) C.A.D.de Kort, Hormones and the structural and biochemical properties of the flight muscles in the Colorado beetle. Ph.D.Thesis, University of Nijmegen (1969).
(38) A.K.M.Bartelink, Synthesis of mitochondrial protein in the flight muscles of the Colorado beetle, Ph.D.Thesis, University of Wageningen (1975).
(39) J.Chudakova, Endocrine regulation of degeneration of cricket (Acheta domestica) deafferented flight muscles. Abstr. comm. Vth.Conf.Eur. Comp.Endocr.Utrecht (1969).
(41) K.H.Trautmann, Evidence of the juvenile hormone methyl (2E, 6E)-10,11-epoxy-3,7,11-trimethyl-2,6-dodecadienoate (JH3) in Insects of four orders, Z.Naturforsch. 29, 757 (1974).
(42) L.Riddiford, Effects of the juvenile hormone on the programming of postembryonic development in eggs of the silkworm, Hyalophora cecropia, Develop.Biol. 22, 249 (1970).
(43) F.Lukoschus, Untersuchungen zur Metamorphose der Honigbiene, Ins.Soc. 2, 147 (1955).
(44) D.J.Wang and R.W.Shuel, Actions of Royal Jelly on honey bee development V. The influence of diet on ovary development, J.apic.Res. 4, 149 (1965).
(45) S.E.Dixon and E.Moser, Duality in function in C.A. of honey bee larvae, Can.J.Zool.50, 593 (1972).

(46) J.Zdárek and O.Haragsim, Action of juvenoïds on metamorphosis of the honey bee, Apis mellifera, J.Insect Physiol. 20, 209 (1974).
(47) F.Lukoschus, Zur Kastendetermination bei der Honigbiene, Z.f.Bienenforsch. 3(8), 190 (1956).
(48) F.Lukoschus, Die Kastenentwicklung bei der Honigbiene:Avitaminose oder endokrine Anaplasie? Verh.Dtsch.Zool.Ges.Hamburg, 278 (1956).
(49) A.Müssbichler, Die Bedeutung äusserer Einflüsse und der Corpora allata bei der Afterweiselentstehung von Apis mellifica, Z.vergl.Physiol. 34, 207 (1952).
(50) E.A.Goewie (publication in preparation).
(51) L.M.Bertholf, The moults of the honeybee J.econ.Ent.18, 280 (1925).
(52) E.A.Goewie (publication in preparation).
(53) P.Wirtz, Differentiation in the honeybee larva, Mededelingen Landbouwhogeschool Wageningen, 73-5 (1973).
(54) J.J.Bounhiol, Recherches expérimentalis sur la déterminisme de la métamorphose chez les Lépidoptères, Bull.biol.Fr.Belg.Suppl. 24,1 (1938).
(55) H.Rembold, Biochemie der Kastenentstehung bei der Honigbiene, Proc.VI Congr. I.U.S.S.I. Bern, 239 (1969).
(56) N.Weaver, Control of dimorphism in the female honeybee, Science, 138, 995 (1962).
(57) N.Weaver, Control of dimorphism in the female honeybee. 4. Techniques for fractioning royal jelly into its major components, J.apic.Res. 13 (2), 103 (1974).
(58) A.Butenandt and H.Rembold, Über den Weiselzellenfuttersaft der Honigbiene. I. Isolierung, Konstitutionsermittlung und Vorkommen der 10-Hydroxy- 2-decensäure. Hoppe-Seylers' Z.f.physiol.Chem. 308, 284 (1957).
(59) A.Butenandt, Wirkstoffe des Insektenreiches, Nova Acta Leopoldina, Barth Verlag, Leipzig (1955).
(60) P.Wirtz and J.Beetsma, Induction of caste differentiation in the honeybee (Apis mellifera L.) by juvenile hormone. Ent.exp.et appl.15, 517 (1972).
(61) H.Rembold and H.Graf, Isolierung und Charakterisierung eines Kastenspezifisches Proteins aus der Honigbiene. Hoppe-Seylers' Z.f. physiol.Chem. 353, 1615 (1972).
(62) S.J.Canetti, R.J.Shuel and S.E.Dixon, Studies in the mode of action of royal jelly in honeybee development. IV. Development within the brain and retrocerebral complex of female honeybee larvae, Can.J.Zool. 42, 229 (1964).
(63) F.Lukoschus, Untersuchungen zur Entwicklung der Kasten-merkmale bei der Honigbiene (Apis mellifica L.), Z.Morph.u.Ökol.Tiere, 45, 157 (1956).
(64) O.Pflugfelder, Volumetrische Untersuchungen an den Corpora Allata der Honigbiene, Apis mellifica L. Biol.Zbl. 67, 223 (1948).
(65) J.Naisse, Contrôle endocrinien de la differenciation sexuelle chez Lampyris noctiluca (Coleoptera, Lampyridae). II. Phenomènes neurosecrétoires et endocrines au cours du développement postembryonnaire chez le mâle et le femelle, Gen.comp.Endocr. 7, 85 (1966).
(66) G.M.Copijn, J.Beetsma and P.Wirtz, Queen differentiation and mortality after application of different juvenile hormone analogues on worker larvae of the honeybee (Apis mellifera L.), (publication in preparation).
(67) A.ten Houten and J.Beetsma, Effects of a juvenile hormone analogue on the development of the hypopharyngeal and wax glands in the adult worker honeybee (Apis mellifera L.)(publication in preparation).

(68) A.S.Johansson, Neurosecretion in the milkweed bug, Oncopeltus fasciatus (Dallas), In: IIth Int.Symp.Neurosecr. Lund, 98 (1958).
(69) J.de Wilde, Reproduction-endocrine control, In: M.Rockstein, ed. The Physiology of Insecta, 1, 59 (1973).
(70) J.de Wilde, Hormones and insect diapause, Mem.Soc.Endocr. 18, 487 (1971).
(71) A.R.McCaffery and K.C.Hignam, Effect of corpus allatum hormone and its mimics on the activity of the cerebral neurosecretory systems of Locusta migratoria migratoroides R and F. Gen.comp.Endocr. 25, 373 (1975).
(72) A.R.McCaffery and K.C.Highnam, Effects of the C.A. on the activity of the cerebral neurosecretory system of Locusta migratoria migratoroïdes R and F. Gen.comp.Endocr. 25, 358 (1975).
(73) J.Jung-Hoffmann, Die Determination von Köningin und Arbeiterin der Honigbiene. Ein Vergleich des Brutpflege massnamen und der Herkunft der Futterkomponenten, Z.Bienenforschung, 8, 296 (1966).
(74) R.W.Shuel and S.E.Dixon, The importance of sugar for the pupation on the worker honeybee, J.apic Res. 7(3), 109 (1968).
(75) Y.Lenskey, The effect of sugar and juvenile hormone on the differentiation of the female honeybee larvae (Apis mellifera L.) to queens, (Publication in preparation).
(76) J.P.Röseler, Unterschiede in der Kastendetermination zwischen den Hummelarten Bombus hypnarum und Bombus terrestris, Z.Natürforsch. 25 (5), 543 (1970).
(77) H.H.W.Velthuis and F.M.Velthuis-Kluppell, Caste differentiation in a stingless bee, Melopona quadrifasciata Lep., influenced by juvenile hormone application, Proc.Kon.Ned.Acad.Wetensch. C 78 (1), 81 (1975).
(78) G.M.Ritcey and S.E.Dixon, Postembryonic development of the endocrine system in the female honeybee castes, Apis mellifera L. Proc.ent. Soc.Ont. 100, 124 (1969).
(79) F.Lukoschus, Das innersekretorische System und seine Bedeutung für die Entwicklung des weiblichen dimorphismus bei der Honigbiene, Symp.Genetica et biol.it.X. Atti IV Congr. U.I.E.I.S. - Pavia,29 (1961).
(80) H.Rembold, Ch.Croppelt and P.J.Rao, Effect of juvenile hormone treatment on caste differentiation in the honeybee, Apis mellifera, J.Insect Physiol. 20, 1193 (1974).
(81) M.Lüscher, Environmental control of juvenile hormone (JH) secretion and caste differentiation in termites, Gen.comp.endocrinology, 3, 509 (1972).
(82) K.Wanyonyi and M.Lüscher, The action of juvenile hormone analogues on caste development in Zootermopsis (Isoptera), Proc. I.U.S.S.I. VIIth Intern.Congr. 392 (1973).

THE ROLE OF DETERMINATOR IN CASTE FORMATION IN THE HONEY BEE

Heinz Rembold
Max-Planck-Institut für Biochemie
D-8033 Martinsried, Germany

ABSTRACT

Endocrine system of queen and worker larvae exhibits marked histological differences in the structure of neurosecretory cells and corpus allatum within first 3 days of larval life. Between 3 - 5 days of larval development, both the castes undergo an endocrine cessation. Endocrine activity returns to normal earlier in the queen larva. - A statistical evaluation of 5 imaginal characteristics (mandibles, basitarsus, ovary, receptaculum seminis, ventral ganglion) shows a tight correlation of shape of mandibles and size of ovaries. Five statistical classes are used to classify adult honey bees, reared under experimental conditions, into workers, intercastes, and queens. - Juvenile hormone is very labile on alkaline hydrolysis, as is demonstrated with the model compound methyl-6,7-epoxy-t-geranoate (I). During hydrolysis the compound adds in a fast reaction either MeOH or water at the epoxide ring, whereas methyl ester is split very slowly. - JH titre is controlled by a degrading system: high enzyme activity and low JH titre are found in worker larval stages L2 - L3 and L5 - L6, whereas stage L4 is characterized by a low enzyme activity and a high JH titre, according to Galleria test. - JH I, JH III or ZR 512, either applied topically or in the food, induce formation of more or less queenlike animals. Correlation analysis shows a rather independent induction of morphological characteristics. This is also apparent from a malformation of the eyes, induced by JH application. - Worker larvae from stages L1 or young L2 (0.2 - 1.5 mg), can be reared on royal jelly in high yield to adults without queen or intercaste formation. Two factors from yeast, a growth and a determination stimulator, have been partially purified. Both are needed for an additional determining activity of royal jelly. - Role of determinator in honey bee caste formation is discussed in connection with these new physiological and chemical data.

INTRODUCTION

The fact, that both queen and worker honey bee castes own the same genom, that they only differ in their phenotype, gives a good chance to the biochemist for studies on gene expression, morphogenesis, and their regulation. Present state of knowledge has been reviewed (Ref 1-7) and much has been speculated, but even after more than fifty years of endeavour, the biological phenomenon neither of honey bee nor of insect caste formation at all, can be understood. From the studies of Zander and Becker (8), who transferred worker larvae of different age into queen cells and followed their subsequent development, we know that the newly hatched larva is sensitive to queen induction till end of third larval instar, i.e. an age of about 3.5 days. We have learnt in the meantime to rear the larvae under controlled conditions in the incubator, a queen bee determining principle was demonstrated in royal jelly, the food of the 3d queen larva (Ref 9), and has been partially purified (Ref 10). But with proceeding purification, unproportional decrease in biological activity was found: there were obviously more factors necessary for queen induction, and some went lost during chromatography. A broad evaluation of endocrine, morphological and nutritional factors became necessary, before further purification of queen bee determinator becomes possible. Our present state of experimental knowledge shall be summarized in the following.

ENDOCRINE SYSTEM OF LARVAE

Results from a comparative study of the endocrine system of queen and worker larvae (Ref 11) suggest that there is a significant difference in the developmental pattern of both the castes. The neurosecretory cells grow faster in the queen and a chiasma is visible around L2 - L3, whereas only from prepupal stage onwards crossing of axons can be seen in workers. No stainable colloids can be found in the neurosecretory cells, if aldehyde fuchsin technique is used, possibly due to a high turnover of neurosecretory material. Between L3 - L5, corpus allatum undergoes a change in the histological picture in both the castes. The neurosecretory cells likewise change their pattern, a little earlier than CA. This cessation of endocrine tissue continues for a period of 36 to 48 hours and then the endocrine activity returns to normal as explicited by nuclear-cytoplasmic relation. Many of queen larvae renew their endocrine activity on the 4th day and thus a little earlier than the worker. Though it is not clear from this histological study, whether neurosecretion is involved in the inhibition of corpus allatum activity, the remarkable fact is, that this type of cellular and of endocrine cessation is visible after the sensitive phase of determination.

MORPHOLOGICAL CHARACTERISTICS

Presence of a queen determining activity in a test food is detected by rearing young worker larvae in the incubator to adults. Only these exhibit characteristics for a queen, an intercaste, or a worker. For a routine test, in which several hundred of adults have to be checked within a few days, shape of mandibles and of hindlegs has been used to characterize determinator activity (Ref 10). But the most important difference between queens and workers is the huge size and high activity of ovaries in the reproductive animals. For this reason, a statistical evaluation was undertaken of five morphological differences: mandibles, basitarsus of the hindleg, diameter of ovaries, diameter of receptaculum seminis, and shape of ventral ganglion (Ref 12). As shown in figure 1, shape of mandible and hairs on inner surface of basitarsus are divided into five statistical classes, and the same was done with the other characteristics. Each class represents 20 % of total. For our problem, only correlation of three parameters is of interest: data for mandible, basitarsus, and ovary are collated in table 1.

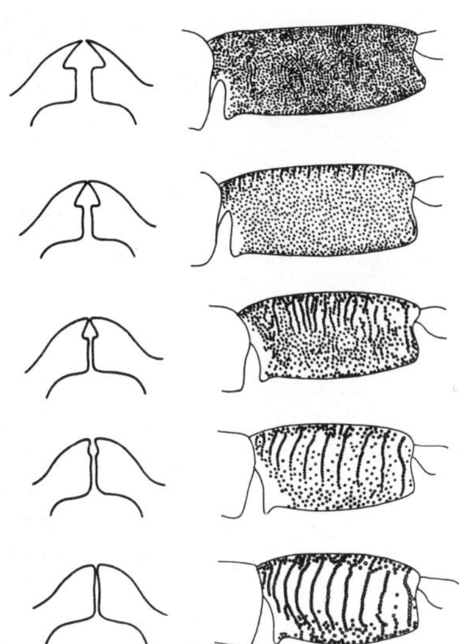

Fig. 1. Scheme for morphological classification of mandibles and inner surface of basitarsus. At the top characteristics of a queen are shown with the notched mandible and evenly distributed hairs on a larger basitarsus. Each of the characteristics equals a statistical weight of 20 %.

TABLE 1: Statistical evaluation of honey bees reared in the incubator, according to differences in mandibles, basitarsus, and ovary (Ref 12).

mandible ± 10 %	basitarsus %	± %	ovary %	± %	individuals n
90	78,0	13,3	75,3	22,0	91
70	73,0	12,7	63,5	13,5	65
50	56,8	10,5	27,3	6,7	44
30	33,3	5,8	12,3	1,9	46
10	1,3	0,0	1,0	0,0	272
basitarsus ± 10 %	**mandible %**	**± %**	**ovary %**	**± %**	**individuals n**
90	96,5	32,1	77,0	21,5	14
70	85,3	35,4	67,0	24,7	149
50	48,0	10,5	26,8	7,3	51
30	22,8	2,7	9,3	1,5	34
10	0,0		1,0	0,0	271
ovary ± 10 %	**mandible %**	**± %**	**basitarsus %**	**± %**	**individuals n**
90	90,8	40,5	75,0	16,0	19
70	89,0	40,9	76,3	14,5	94
50	72,3	24,2	67,5	16,6	37
30	42,3	13,8	45,8	15,0	29
10	5,8	0,7	6,8	0,4	255

According to table 1, a good correlation exists between external caste specific characteristics like shape of mandible or hindleg, and size of ovary. If, for example, shape of mandible falls into class 70±10 % (compare Fig. 1), basitarsus and ovary of the same animal are with about identical probability in the same class. Vice versa, an ovary of big size is always coupled with a more or less notched mandible. For our routine test, we characterize animals with strongly notched mandibles (class 60-100 %) as queens, the following two classes as intercastes (class 30±10 % only, if more than 20 ovarioles per ovary), and class 10±10 % as workers.

JUVENILE HORMONE AND CASTE FORMATION

Chemical aspects

For preparation of juvenoate, juvenile hormone is usually hydrolized with 0.5N NaOH in methanol/water (1:1) to the corresponding acid (Ref 13-15). High reactivity of the epoxide ring could interfere with such a simple ester hydrolysis. A kinetic study with methyl 6,7-epoxy-t-geranoate (I) as a model compound proved indeed, that in the course of hydrolysis the epoxide ring is attacked in a fast, base catalysed reaction, and either methanol is added to the methyl ether II or water to the diol ester III. This reaction is quantitative at room temperature within a few minutes, whereas ester hydrolysis proceeds in a very slow reaction within about 48 hours according to Fig. 2.

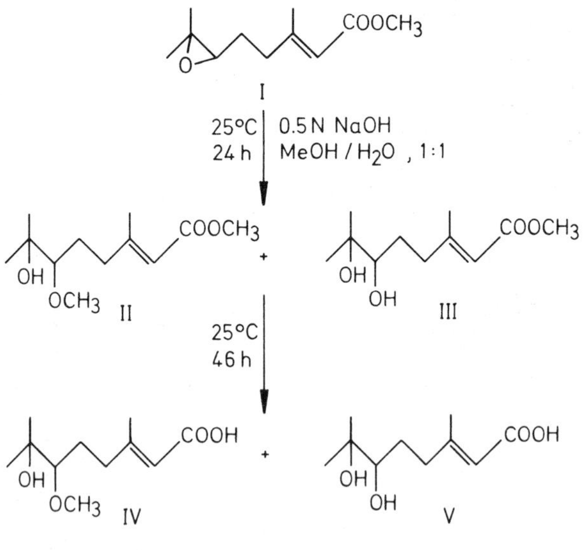

COOCH3
O
I
25°C 24 h | 0.5 N NaOH MeOH / H_2O , 1:1
COOCH3 + COOCH3
OH OCH3 II
OH OH III
25°C 46 h
COOH + COOH
OH OCH3 IV
OH OH V

Fig. 2. Reaction scheme for hydrolysis of methyl 6,7-epoxy-t-geranoate with 0.5N NaOH in $MeOH/H_2O$ (1:1).

Turnover of juvenile hormone in bee larvae

From histological studies (Ref 11) it is obvious, that size and possibly activity of corpora allata is coupled with caste formation in the honey bee. Studies on JH degrading activity in worker larvae (Ref 16) and JH activity (Ref 17) revealed a clear correlation of degrading activity and of JH titre (Fig. 3). If JH degrading and general esterase activity are compared it is obvious, that JH turnover is under control of a specific enzyme, the activity of which is high during first three days of larval development, a period which is almost identical with the sensitive phase for caste induction, and again during end of active food uptake. Very low JH degrading activity is found around the 4 day larva and in the pharate pupae. From several stages, juvenile hormones were extracted and quantified, using isotope dilution technique and Galleria test. Our present data, as collated in Fig. 3, speak in favour of a direct control of JH titre by one or more inactivating enzymes, one of which seems to be a specific esterase. Remarkably enough, degrading activity in haemolymph is almost zero in young larval stages and relatively low also in the 5 and 6d old larva. It is highly probable, that most of JH will be inactivated in the fat body. The same studies with queen larvae have now been started.

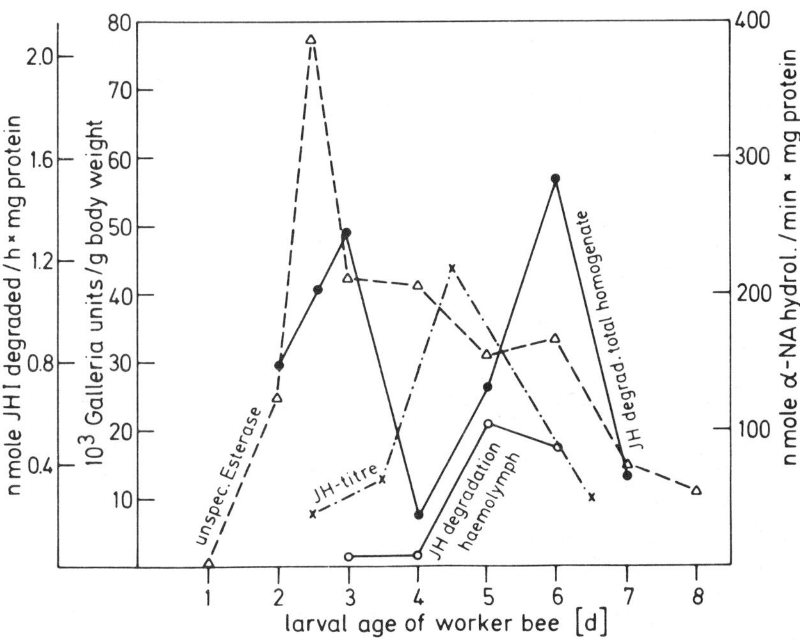

Fig. 3. JH I degrading capacity of total homogenates and haemolymph from different worker larval stages of the honey bee. For comparison, general esterase activity of the same extracts is measured with a-naphthyl acetate as a substrate (Ref 16). For Galleria test, lipids were partially purified by TLC, using isotope dilution technique for quantitative calculation (Ref 17).

TABLE 2. Statistical evaluation of adults reared from 2d larvae in the incubator. One group of the larvae was treated with juvenile hormones (µg/animal), the other reared untreated. For more details compare table 1 (Ref 12).

	Control (untreated)					Juvenile hormone treated				
mandible ± 10 %	basitarsus %	± %	ovary %	± %	individuals n	basitarsus %	± %	ovary %	± %	individuals n
90	93,8	37,5	75,0	0,0	4	100,0	0,0	100,0	0,0	1
70	73,5	8,9	50,0	17,2	16	78,0	20,1	51,5	20,0	17
50	57,3	9,9	28,8	7,7	14	51,3	9,5	41,8	9,5	24
30	35,0	6,1	19,8	4,2	15	16,8	2,3	26,8	3,8	36
10	0,8	0,02	1,0	0,03	90	0,5	0,01	5,3	0,05	91
basitarsus ± 10 %	mandible %	± %	ovary %	± %	individuals n	mandible %	± %	ovary %	± %	individuals n
90	100,0	0,0	75,0	0,0	3	81,3	23,2	62,5	43,3	4
70	69,3	19,3	51,3	16,9	21	69,3	13,5	54,5	14,0	17
50	45,0	7,9	21,8	4,8	15	48,8	6,3	39,0	8,0	20
30	21,0	2,0	13,8	2,5	12	27,0	2,8	24,0	3,8	25
10	0,0	0,0	1,0	0,03	90	2,8	0,2	9,0	1,1	91
ovary ± 10 %	mandible %	± %	basitarsus %	± %	individuals n	mandible %	± %	basitarsus %	± %	individuals n
90		-		-	0	100,0	0,0	100,0	0,0	1
70	82,0	30,7	82,0	22,2	11	70,0	20,0	78,0	24,8	10
50	53,5	15,6	60,0	17,0	15	43,0	12,3	46,0	15,5	25
30	34,3	10,5	38,8	13,6	11	21,3	4,4	15,8	3,2	59
10	6,5	0,9	7,5	1,1	90	3,5	0,4	3,3	0,3	91

Effect of JH on larval development

Some speculations have been made that juvenile hormone is directly involved in honey bee caste formation (Ref 18,19), but due to our own data, topical application of JH I to 2d larvae is followed only by a shift from worker to intercaste production. To a high extent the adults exhibited malformations. Mainly the eyes showed severe alterations with characteristics of a somatic mutation (Ref 20). It is to be expected, that application of juvenile hormone in the range of micrograms per animal will severely influence morphogenesis. A series of adults, either coming from untreated or from JH treated larvae, were classified according to table 1. Data collated in table 2, clearly prove that correlation of external characteristics like mandible or basitarsus, and size of ovaries is less with JH treated animals than with untreated. Main effect of JH application on young worker larvae is a stimulation of ovary development in the adults.

TABLE 3. Effect of JH I, JH III, and ZR 512 (Zoecon) on caste formation in the honey bee. 2d Larvae were reared on basic food (Ref 10), to which 60 μg of hormone was mixed, corresponding to 1 μg per larva (Ref 12).

hormone added	adults n	queens n	queens %	intercastes n	intercastes %	workers n	workers %	determination %
-	21	0	0	0	0	21	100	0
JH I	28	3	11	19	68	6	21	79
JH III	36	0	0	7	19	29	81	19
ZR 512	30	0	0	10	33	20	66	33

As shown in table 3, JH I has the highest effect on bee larvae, followed by the mimic ZR 512 and JH III with considerably less activity. With all the three hormones, formation of intercastes is induced, but the queens grown after feeding JH I showed another correlation of queen characteristics than queens grown on royal jelly.

As mentioned before, the JH treated larvae develop in a high yield to adults with eye malformations. A histological study (Ref 21) revealed, that also in the malformed eye all cell groups are present, which build up the ommatidium, but their regular arrangement is missing. Mainly the cornea is altered (Fig. 4), showing an irregular structure instead of the regular facets in the compound eye. The whole histological picture of the malformed area shows a lack in organisation during morphogenesis, which must have been induced during JH treatment in the imaginal disc already. An electronmicroscopic view (Fig. 5) of a malformed eye shows a clearly defined border between normal and altered structures. These malformations are always directed to the forehead and differ from a small sickle to a complete alteration of both the eyes. Interesting enough, outer border or size of the eye are never influenced.

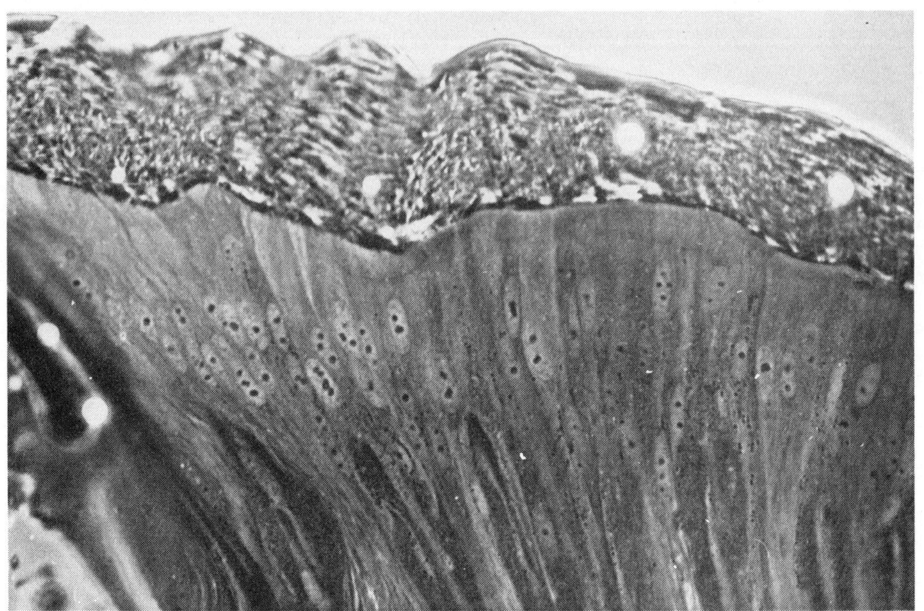

Figure 4

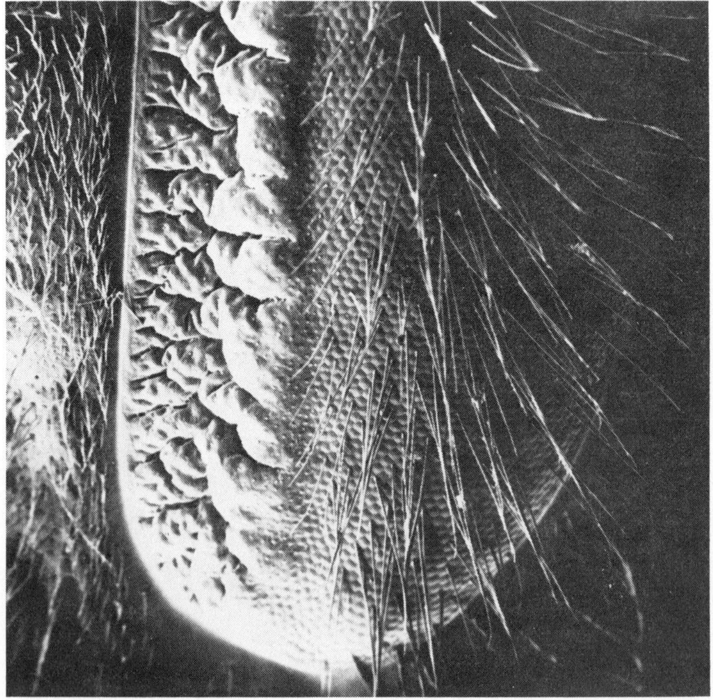

Figure 5

Fig. 4. Longitudinal section through area of malformation. Thickened cornea and nuclei of outer pigment cells are localised in a distal position in the ommatidia. Magn. 550 x (Ref 21).
Fig. 5. Lower part of a honey bee eye with malformation. Forehead on the left side. Magn. 70 x (Ref 21).

NUTRITIONAL FACTORS

Royal jelly as collected from queen cells of 3 day larvae is a semisolid mass which must be liquefied for use in a biological test. We usually dilute with a mixture of glucose, fructose, and yeast extract in water, to which for dilution of basic food vitamines and some amino acids are added (Ref 10). One can easily imagine that by such a procedure some essential components could be diluted under a critical concentration and now limit growth, sensitivity against determination, or influence the endocrine system. As it has not yet been possible to prepare a synthetic diet, one has to use an indirect way in order to supplement such compounds which have either been partially destroyed during storage or the concentration of which is below a critical level. Table 4 presents data which prove the existence of two factors in yeast extract, one of which is essential for growth stimulation, and another one necessary for a response of the larvae to queen bee determinator.

TABLE 4. Effect of yeast extract on queen bee formation. A mixture of 20 g royal jelly and 20 ml of a solution of each 2.5 g D-glucose and D-fructose in water is the control, to which either yeast extract or a chromatographic fraction from the corresponding amount of yeast extract was added (Ref 22).

factors added	adults n	queens n	intercastes n	workers n	determination %
–	41	0	5	36	12
0.5 g yeast extract	49	2	10	37	24
Bio-Gel fraction 11-43 (1g)	49	17	4	28	43
Bio-Gel fractions 11-29 (1g)	7	0	0	7	0
Bio-Gel fractions 30-43 (1g)	38	3	2	33	13

The results indicate that in yeast extract an activity is contained, which is necessary for queen bee induction. The fact that yeast extract added to basic food stimulates only to a rather low extent, shows that this effect is not identical with the determinator. But the factors must be present to enable determination via a sensibilisation of the larva. Using gel chromatography on Bio-Gel P2, two factors could be separated from each other (compare table 4). The one, contained in fractions 11-29, is for itself rather toxic, the other one (fractions 30-43) obviously stimulates food uptake but influences neither rate of survival nor of determination. If both are put together as the same aliquot of

TABLE 5. Development of honey bee larvae reared on royal jelly without (upper part) and with addition of 1 g yeast extract (fraction 11-43). For further details compare table 4. Classification of adults according to table 1: M1 = mandible class 80 - 100 %, M5 = 0 - 20 %, M4: 2/1 means, that 2 adults have more, 1 less than 20 ovarioles (Ref 22).

days	1	2	3	4	5	6	7	8	9	10	11	12	13	14	15	16	17	18	remarks
larvae	60	60	60	60	57	55	54	53	45	3									food without
pupae								1	6	40	43	43	43	43	40	37	21	2	yeast extract.
adults															3	3	16	19	12 % determi-
differentiation											M1								nation
											M2								
											M3				1				
											M4				2	2/1			
											M5						16	19	
larvae	60	60	60	59	59	59	57	41	17	7	6								food with
pupae							1	17	24	50	51	48	39	31	26	24	9		yeast extract.
adults												1	9	8	5	2	15	9	43 % determi-
differentiation											M1	1	9	3					nation
											M2			3					
											M3			1	1				
											M4				2/2	-/2			
											M5						15	9	

1 g yeast extract (fractions 11-43), they remarkably stimulate queen formation. Both factors have been partially purified: the first one is contained in the fraction of amino acids, the second is eluted at the end of chromatography together with inorganic salts.

If nutrients are necessary for queen bee induction, one could speculate that the key for an understanding of caste formation is just a problem of nutritional balance. Extensive comparative analyses do not support this assumption (Ref 3), but without a completely synthetic diet the possibility of limiting traces in the food cannot be excluded. A careful observation of imaginal development, as given in table 5 for the both examples of a royal jelly without and with yeast extract, shows some important facts:

1. Development of the castes followes a "developmental clock" also under experimental conditions. Larvae becoming queens also pupate first, then follow the intercastes and then the workers.

2. Yeast extract induces formation of strong queens. Most of the 18 pupae from days 7 and 8 developed to queens in mandibular class M1 (80 - 100 %).

3. Without yeast extract, duration of development is not influenced. Pupae from days 8 and 9 consequently developed to adults of mandibular classes M3 and M4, where also adults from the test with yeast extract hatched.

4. As collated in table 6, body weight of newly hatched adults is not related with determination. Under experimental conditions, average body weight of queens and intercastes is about the same as that of workers.

TABLE 6. Body weight (mg) of newly emerged adult honey bees under experimental conditions (Ref 22).

Queens + Intercastes	Workers
129, 148, 150, 156, 157, 160, 160, 167, 167, 169, 185, 185, 188, 194	124, 140, 151, 165, 175, 176, 179, 182, 184, 186, 188, 193, 194, 197
Ø : 165	Ø : 173

ROLE OF DETERMINATOR

Many theories have tried to explain induction of caste formation in the honey bee, depending on the experimental knowledge (for review see Ref 1-7). Let us go through those, which can be discussed with regard to our recent results.

1. Caste formation is induced by a different amount of food given to the young bee larva. A queen bee determinator could then have the function of a feeding stimulant. This hypothesis bases on the two facts that queen larvae are fed a surplus of food in a conti-

nous flow, whereas worker larvae receive less food, and that body weight of a newly hatched queen is around 200 mg, of a worker only about 100 mg. But under experimental conditions all the adults have a body weight of around 160 mg, queens and intercastes as well as workers. This applies to larvae grown on an inactive food as well as on a food with high determining activity. The same weight of all the castes and intercastes can be explained by the fact, that all larvae receive the same amount of surplus food and grow in a vessel, where growth is not limited by lack of space as in the narrow worker cell. To sum up: queen bee determinator has not the function of a feeding stimulant.

2. After the sensitive developmental phase, i.e., after about third day of larval development, worker larvae subsequently receive a mixture of pollen and honey, whereas queen larvae are continously fed with a mixture of glandular secretion and honey. This food change could influence endocrine activity, visible in the histological picture of both neurosecretory cells and corpora allata, and by that way switch off development of ovaries with all consequences. Against that assumption speak our experimental results: worker larvae develop with the same food either to workers or to queens. Difference in food quality of the elder larvae does not trigger development of workers.

3. Caste formation is a problem of nutritional balance. Jung-Hoffmann (23) observed that bee larvae receive two different secretions from their nurses, a transparent and an opaque, the proportion of which is altered if queen or worker larvae are fed. Also the amount of each food differs considerably: queen larvae are fed about 1600 times, worker larvae only 143 times. If we for example assume, that some chemically labile food components are inactivated very fast, one can easily imagine a qualitative difference between worker and queen food in the cell already, which is a quantitative one in practice. Royal jelly, as collected and stored for a longer time, would then have lost its activity according to some limiting nutrients, which can be added through yeast extract for example in order to restore nutritional balance. Determining quality of carefully collected royal jelly sometimes differs considerably, if used in our feeding test, though determinator activity is present. We can sum up: nutritional balance is a prerequisite for caste induction. Only then determinator can switch on "queen programme".

This third assumption is supported by our present knowledge. It also explains, why further purification of queen determinator is blocked without an optimal nutritional balance in the test food. If that problem is not solved, we lose determining activity just because we separate the determinator from an impurity, which is necessary for nutritional balance. Our present endeavour is to separate the growth stimulating and the determination supporting factors from yeast extract in order to add them in their appropriate proportion to our basic food. This is necessary, because complete yeast extract in higher concentrations severely interferes with larval growth mainly from its high content of amino acids and other nitrogen containing compounds. Such a basic food must provide an optimal nutritional balance for larval growth combined with minimal determining activity. Role of determinator is then to switch on as a primum movens the whole network of regulatory events like activation of endocrine system, of timing and other morphogenetic genes, which program queen bee development.

ACKNOWLEDGEMENT

Our studies collated above, have been supported by a Grant, BCT 86, by Bundesministerium für Forschung und Technologie.

REFERENCES

(1) G.F. Townsend, R.W. Shuel, Some recent advances in apicultural research. Ann. Rev. Entomol. 7, 481-499 (1962).
(2) H. Rembold, Die Kastenentstehung bei der Honigbiene, Apis mellifera L. Naturwiss. 51, 49-54 (1964).
(3) H. Rembold, Biologically active substances in royal jelly. Vitamins and Hormones 23, 359-382 (1965).
(4) H. Rembold, Biochemie der Kastenentstehung bei der Honigbiene. Proc. VI Congr. IUSSI, Bern, 239-246 (1969).
(5) H. Rembold, Die Kastenbildung bei der Honigbiene, Apis mellifera L., aus biochemischer Sicht. G.H. Schmidt, Ed., Sozialpolymorphismus bei Insekten. Wissenschaftl. Verlagsgesellschaft, Stuttgart, 694-739 (1974).
(6) H. Rembold, Steps in honey bee caste determination. Proc. symp. pheromones and defensive secretions in social insects, Dijon, 189-195 (1975).
(7) N. Weaver, Physiology of caste determination. Ann. Rev. Entomol. 11, 79-102 (1966).
(8) E. Zander, F. Becker, Die Ausbildung des Geschlechtes bei der Honigbiene. Erlanger Jb. Bienenk. 3, 163-223 (1925).
(9) H. Rembold, G. Hanser, Nachweis des determinierenden Prinzips im Futtersaft der Königinnenlarven. Z. physiol. Chem. 339, 251-254 (1964).
(10) H. Rembold, B. Lackner, I. Geistbeck, The chemical basis of honeybee, Apis mellifera, caste formation. Partial purification of queen bee determinator from royal jelly. J. Insect Physiol. 20, 307-314 (1974).
(11) G.S. Dogra, H. Rembold, A comparative study of the endocrine system of the honey bee larvae under normal and experimental conditions. Z. Naturforsch. 1976, subm. for publ.
(12) H. Rembold, Ch. Czoppelt, Effect of juvenile hormones on caste formation in the honey bee. In preparation.
(13) M. Slade, Ch.H. Zibitt, Metabolism of Cecropia juvenile hormone in insects and in mammals. J.J. Menn, M. Beroza, Ed., Insect Juvenile Hormones. Acad. Press, New York, 155-176 (1972).
(14) A.F. White, Metabolism of the juvenile hormone analogue methyl farnesoate 10,11-epoxide in two insect species. Life Sci. 11/II, 201-210 (1972).
(15) R.C. Lauer, P.H. Solomon, K. Nakanishi, B.F. Erlanger, Antibodies to insect C_{16}-juvenile hormone. Experientia 30, 558-560 (1973).
(16) S.D. Mane, H. Rembold, Juvenile hormone degrading activity in worker larval stages of the honey bee. In preparation.
(17) H. Rembold, H. Hagenguth, M. Lüscher, Juvenile hormone titres during metamorphosis of honey bee castes. In preparation.

(18) P. Wirtz, J. Beetsma, Induction of caste differentiation in the honeybee (Apis mellifera) by juvenile hormone. Entomologia exp. appl. 15, 517-520 (1972).
(19) P. Wirtz, Differentiation in the honeybee larva. Commun. Agric. Univ. Wageningen 73-5, 1-155 (1973).
(20) H. Rembold, Ch. Czoppelt, P.J. Rao, Effect of juvenile hormone treatment on caste differentiation in the honeybee, Apis mellifera. J. Insect Physiol. 20, 1193-1202 (1974).
(21) P. Seifert, Zur Histologie einer durch den Einfluß von Juvenilhormon hervorgerufenen Mißbildung am Komplexauge der Honigbiene, Apis mellifera L. Diplomarbeit, Univ. München, 1976.
(22) H. Rembold, B. Lackner, Nutritional factors which influence caste formation in the honey bee. In preparation.
(23) I. Jung-Hoffmann, Die Determination von Königin und Arbeiterin der Honigbiene. Z. Bienenforsch. 8, 296-322 (1966).

ENVIRONMENTAL, GENETIC AND ENDOCRINE INFLUENCES IN STINGLESS BEE CASTE DETERMINATION

H.H.W. Velthuis
Laboratory of Comparative Physiology, Utrecht, The Netherlands.

INTRODUCTION

The stingless honeybees are a vast group of social bees living in the tropics. Systematically, the group can be divided into the Meliponini and the Trigonini. While the first group is rather homogenous and restricted to the neotropics, the Trigonini are rather diverse, ranging from the most primitive to the most evolved social structures and with a worldwide tropical distribution.

In precolumbian times several species, both from the genus *Melipona* and from the subgenus *Scaptotrigona* were domesticated by the indians of Central and South America because of their honey (see, for instance, Wagner 1960). This domestication is still in use and the honey is highly appreciated. Nogueira-Neto (1970) wrote a concise manual for Meliponoculture.

Although the Meliponids are the closest relatives of the honeybees of the genus *Apis* and their most evolved species can be regarded as ranking as high in their social development, there are a number of characteristics in which the group shows an alternative evolution. For instance, they have a horizontally arranged comb (except in the most primitive species) and a mass provisioning. While in *Apis* the high reproductive capacity of the queen is achieved by an enormous increase in the number of ovarioles (from the basic Hymenopteran number 4 to about 180), in the Meliponids this capacity is reached by a not less considerable elongation of the ovarioles, their number being only 8. Such differences show that in the evolution of the honeybees the Meliponinae and the Apinae separated at an early stage of social evolution. The reader is referred to the synopsis given by Michener (1974) for further systematic and biological information.

The mechanism of caste determination in the genus *Melipona* differs from the trophic determination found in the allied Trigonini and the other social bees. While in the other bees the food quantity and at least in the honeybee also the food quality determines the development into a queen, in *Melipona* a genetic determination has been described (Kerr 1950, Kerr, Stort and Montenegro 1966). In this genus a certain but small variation exists in the quantity of food available for each larva and below a certain amount only workers emerge. With relatively high amounts of food both workers and queens are produced. In such a situation 25 % of queens are found on the average. This led Kerr (1950) to suppose that the female larvae were not identical in their potencies. He envisaged the existence of two queen determining genes, each having two alleles, which are effective only in the heterozygotic condition. The hypothesis has been elaborated in subsequent papers, the latest by Kerr (1975a, 1975b) and Kerr, Akahira and Camargo (1975).

Recently, Darchen and Delage-Darchen (1974, 1975) questioned the genetic determination of *Melipona* and put forward a hypothesis of a queen determining

Fig. 1. Brood nests of some stingless bees.

A. Part of the brood area of *Leurotrigona muelleri*. These minute bees construct irregular clusters of brood cells.

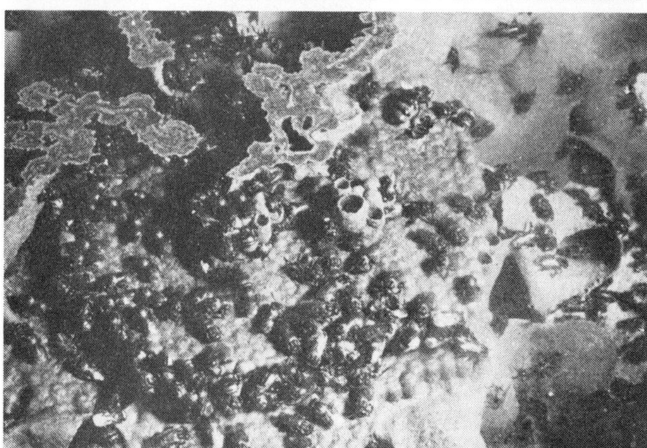

B. *Scaptrigona postica* in an observation hive. Three worker cells and one queen cell are under construction. In the right top corner the queen is leaving the comb (photo by courtesy of Dr. W. Engels).

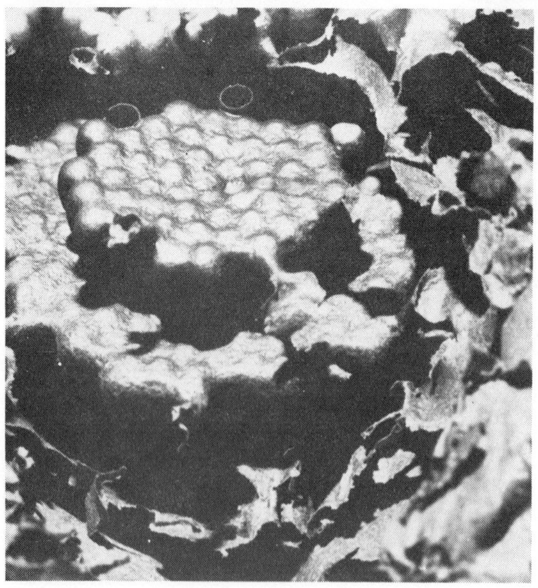

C. Some combs of *Melipona quadrifasciata* within the heavy involucrum. One cell awaits provisioning, the other open cell is not yet completed.

substance admixed to the food, thus bringing the genus *Melipona* back from their unique position into the row of other social bees.

Egg laying and brood development

The brood cells are generally arranged in a number of horizontal combs to the margin of which new cells are added (Fig. 1). These newly built cells are open at the top. The oviposition process, described for a number of species by Sakagami, Zucchi and collaborators since 1963, is a rather complicated behavioural process in which workers and the queen are involved. In some species one cell at a time is provisioned, in others cell provisioning takes place in batches of cells (Sakagami and Zucchi 1974). Immediately after the cell provisioning, which involves the deposition of the content of the honey sac and the food glands of a number of workers into the cell, the queen lays an egg on top of the food and the cell is closed by a worker. The only care subsequently given to the cell is temperature control.

In the more primitive species the cells are arranged in threedimensional irregular clusters, the cells being interspaced by bars of wax permitting the passage of the adults. They have an oviposition process not unlike the other species (Sakagami and Zucchi 1974).

The oviposition process in its generalized form can be regarded as a behavioural chain in which at least the following elements are relevant:

a. inspection of the completed cell by the queen;
b. the subsequent release of a signal during 'cell fixation' of the queen, stimulating the workers
c. to insert their head into the cell and to regurgitate food.
d. In some species between the various food discharges, but at least in all species at the end of the depositions, the queen inserts her head, sometimes coupled with food uptake, before she lays her egg. Then follows
e. immediate cell closure by a worker. Before doing so, in some species and in some periods of the year, this worker bee may lay a fertile egg in the cell, which will become a male (Beig 1972).

Interruption of this chain of events in any stage once food deposition has started leads to removal of all the material in the cell, indicating the highly interactive nature of the oviposition process. Competition between females, in a more or less ritualized form, is a rather distinct aspect of this process (Sakagami, Camilo and Zucchi 1973).

In the genus *Trigona* queen cells can be distinguished from worker cells by their very different volume. In the genus *Melipona*, however, all cells are of the same size and queens emerge randomly from the combs. This already is an indication that where in *Trigona* one of the environmental factors, viz. food quantity may be decisive, in *Melipona* other mechanisms must regulate caste differentiation.

The mass provisioning and the subsequent lack of nursing make the stingless bees excellent animals for experimental work on bee development. C.A. de Camargo (1972a, 1972b) developed the technique for *in vitro* rearing of the immature stages. This procedure involves food transfer from natural cells to artificial wax cells, after which a young larva is grafted onto the food. The cell is closed and incubated at an appropriate temperature and humidity.

CASTE DETERMINATION IN TRIGONA

Two studies on this subject have been undertaken, one by the Darchens (Darchen and Delage 1970; Darchen and Delage-Darchen 1971) on some African species (*Hypotrigona*, *Liotrigona*, *Axestotrigona* and *Dactylurina*) and another by Camargo (1972) on the Brazilian *Scaptotrigona postica*.

For all four African species the Darchens report that transfer of a fullgrown worker larva to newly provisioned cells resulted in a continuation of food intake and finally into a queen adult. This also holds when the additional food came from another species. Terrada (1973) observed that in *Leurotrigona muelleri*, a species with cells in the cluster arrangement, sometimes two cells were not interspaced; the older larva was then able to pierce the common wall and to feed from the neighbouring cell. In this way in this species queens arise; like in the experiments of the Darchens a doubling of the food suffices.

The Darchens also succeeded in rearing a worker by limiting food uptake of a larva from a queen cell of *Axestotrigona eburnensis*.

The Darchens classify their experimentally obtained animals into normal workers normal queens, miniature workers and miniature queens. The last two categories are considered to be the result of undernourishment. Although morphometric data of these animals deviate markedly from those of normal workers and queens, the authors are not tempted to classify them as intercastes. Table 1 gives such a comparison for *Axestotrigona eburnensis*.

TABLE 1 Body proportions of adults of *Axestotrigona eburnensis* reared on unnatural amounts of food. After Darchen and Delage-Darchen 1971.

		$\frac{\text{width}}{\text{length}}$ tibia 3	$\frac{\text{length of eye}}{\text{head width}}$
normal workers, mean		0.45	0.33
miniature worker	1	0.43	0.29
	2	0.46	0.33
	3	0.40	0.34
	4	0.46	0.35
miniature queen	1	0.38	0.46
	2	0.33	0.49
	3	0.31	?
normal queen	1	0.33	0.50
	2	0.33	0.47

At the first sight the situation in the Brazilian *Scaptotrigona postica* seems somewhat different. C.A. de Camargo (1972a + b) reared larvae *in vitro* on different amounts of worker food. A first experiment is given in Table 2. Mortality is related to food quantity; i.e. those larvae that did not consume all of their high amount of food died in their cell.

TABLE 2 Effect of overnourishment on caste development in *Scaptotrigona postica*. In series 1-4 young larvae were transferred to artificial cells containing the food taken from 2, 3, 4 or 5 worker cells resp. In series 5 fullgrown larvae were transferred to a new cell containing the food quantity of one cell. All series started with 40 larvae. The natural weight of a worker is 19.0 ± 0.4 mg, that of an emerging queen 44.5 ± 2.5 mg. After Camargo 1972a.

series	food from	emerged	number	mean weight, mg.
1	2 cells	workers	?	slightly heavier than normal
2	3 cells	queens	9	32.6 ± 0.7
3	4 cells	queens	6	36.0 ± 3.5
4	5 cells	queens	1	45.5
5	2 cells	all died		

TABLE 3 Effect of different food quantities on caste differentiation in *Scaptotrigona postica*. Food and larvae were taken from worker cells. The worker cell contains on the average 35.4 mg. After Camargo 1972a.

food quantity	number of emerged		
(mg.)	workers	intermediates	queens
35.4	10		
70	5	2	3
80	4	3	7
90	1	1	3
100			2
110			2

In a second experiment the critical range between the food quantity from two and three cells was narrowed down (Table 3). With food quantities lower than twice the normal worker quantity the majority of larvae develop into workers, generally taller than the normal ones, while with food quantities higher than 2.5 times the normal worker ration almost only perfect queens emerge.

From the same Table we see that with 2.0 to 2.5 times the natural worker food quantity some of the emerging adults have been classified as intermediates. Table 4 shows to what extent these animals deserve this denomination. Apparently, their most intermediate character lies in the reproductive apparatus. Therefore, the difference in conclusion between the Darchens and Camargo concerning the appearance of intermediates could well be a matter of semantics.

This seems to be the appropriate stage at which to discuss the use of the terms intermediate and intercaste. A number of morphological differences between queen and worker can be expressed numerically. Other aspects of caste differences can be expressed only in a qualitative sense, such as showing a caste specific behaviour or not, to become mated, to release court behaviour etc. Since the principal difference between a worker and a queen lies in their reproductive function, I suggest to relate the use of 'caste' to this function.

TABLE 4 External and internal characteristics of the intercastes of *Scaptotrigona postica*, obtained by Camargo 1972.

			head length		head width		between eyes		length of eye	
		n	mean	range	mean	range	mean	range	mean	range
natural workers			2.14	± 0.01	2.61	± 0.01	1.82	± 0.01	1.50	± 0.01
experimental workers		9	2.32	2.08-2.68	2.69	2.12-2.96	2.00	1.84-2.12	1.56	1.24-1.72
experimental queens		17	2.18	2.08-2.28	2.52	2.40-2.60	1.76	1.40-2.00	1.36	1.32-1.40
intermediates obtained with										
70 mg	1		2.00		2.44		1.76		1.32	
	2		2.04		2.48		1.68		1.36	
80 mg	1		2.16		2.76		1.96		1.36	
	2		2.20		2.60		1.76		1.36	
	3		2.16		2.44		1.84		1.46	
90 mg	1		2.20		2.56		1.80		1.36	

			length of corbicula		width of spermatheca		remarks on ovaries
		n	mean	range	mean	range	
natural workers			2.11	± 0.02	0.10	± 0.01	short, straight
experimental workers		9	2.11	1.96-2.28	0.11	0.06-0.16	short, straight
experimental queens		17	2.31	2.20-2.56	0.61	0.42-0.76	long, spiralized
intermediates obtained with							
70 mg	1		2.08		0.48		long, straight
	2		2.52		0.56		one queenlike, other long, straight
80 mg	1		2.16		0.48		long, spiralized
	2		2.48		0.64		long, spiralized
	3		2.52		0.30		long, spiralized
90 mg	1		2.36		0.58		intermediate

An intercaste would be an animal that shows morphological or behavioural properties characteristic for one of the two natural classes while at the same time other aspects resemble the other natural extreme. On the other hand, those animals that are clearly workers or queens, but in a numerical sense deviate from the naturally occurring variation within each caste, should be denoted intermediates, or superqueens, or superworkers respectively. Thus it might be possible to call an animal an intermediate queen, when she looks like and functions as a queen but in a numerical sense she is not a natural one. In the intercaste conflicting expressions are found; the animal is partly worker-like, partly queenlike. Such morphs are in a sense comparable to mosaic patterns and gynandromorphs, where also different genetic and other molecular information is used in the differentiation of the various parts of the body.

GENETIC OR ENVIRONMENTAL INFLUENCE ON CASTE DIFFERENTIATION IN MELIPONA?

The well-known genetic model of Kerr on caste differentiation in *Melipona* was envisaged as early as 1946. Up to very recently it has been impossible to prove the genetic aspect in a direct way, that is by the study of the progeny of selected crossings. Therefore, all the information available is of an indirect nature and the discussions on the validity of the theory suffer from the fact that the participants belong to different disciplines of biology.

The hypothesis is principally based on the fact that in many cases 25 % of the offspring emerge as queens provided food conditions are optimal. In those cases where queen production is lower environmental factors can be shown to be responsible for such a reduction. Table 5 gives data on queen production in a number of species.

TABLE 5 Queen-worker ratios observed in different *Melipona* species under natural conditions. After Kerr 1974.

species	number of workers	and queens	percentage queens
M. marginata	177	62	25.9
M. quadrifasciata	318	110	25.7
M. nigra	11	4	26.7
M. rufiventris	73	24	24.7
M. interrupta fasciculata	27	7	20.6
M. pseudocentris	142	39	21.5
M. quinquefasciata	22	9	29.0
M. flavipennis	38	3	7.3

Kerr, Stort and Montenegro (1966) studied a number of environmental factors in relation to their effect on *M. quadrifasciata anthidioides*. They observed no effect of temperature fluctuation, feeding by the queen or the number of eggs laid in a single cell (sometimes more than one egg is found; then cannibalism occurs and only one larva survives and gets all the food). However, factors related to the amount of food deposited and consumed are correlated with the percentage of emerging queens. These mutually dependent factors are:

a. the number of workers involved in food deposition; b. the amount of food deposited; and c. pupal weight.

The amount of food deposited in a cell is inversely correlated with the number of workers that are involved in the deposition act. The authors suppose that under favourable food conditions the workers have ample material in their honey stomach and food glands, and therefore the minimum amount of food necessary to release egg deposition by the queen is easily surpassed. Under less favourable conditions each worker adds smaller amounts and the minimum is reached with smaller steps. Our own observations on *M. favosa* show that during the food deposition phase sometimes workers visit the same cell more than once and behave as if each time they deposit food; we may suppose, however, that once the honey stomach is emptied, it will take a relatively long time to refill it. The repeated participation is more frequent when fewer bees react to the releasing stimulus emanating from the queen. Thus this behaviour has the same net effect as has been attributed to the food conditions in the colony by Kerr *et al.* (1966).

Since generally all the food from the cell will be consumed by the larva, pupal weight is correlated with the amount of food deposited. The relation between pupal weight and occurrence of queens is given in Fig. 2. In this graph 252 workers and 50 queens are represented. Below 70 mg of pupal weight 27 workers and no queens were observed. From 70-86 mg 96 worker and 25 queen pupae were found and above 86 mg 133 workers combined with the other 25 queens. Since metabolism in queen pupae is more intense than in worker pupae these weight classes do not fully refer to identical food quantity classes. The mean weight of emerging queens is about 60 mg, of workers 70 mg (Kerr *et al.* 1966).

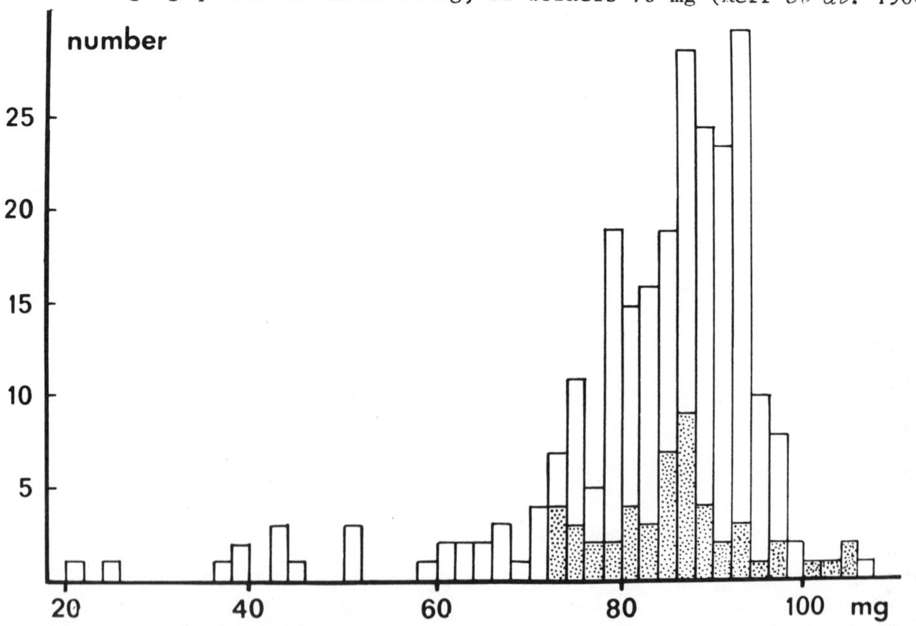

Fig. 2. The relation between pupal weight and occurrence of queens in *Melipona quadrifasciata* (after Kerr, Stort and Montenegro 1966). In the original paper a printing error apparently has been overlooked and has been repeated subsequently: instead of 202 the figure depicts 252 workers. Dotted: queen pupae; open: worker pupae.

The authors conclude from these and similar data that the food quantity plays a decisive role in the expression of queen characteristics; above the critical level a 3:1 worker to queen segregation on the basis of the double heterozygocy occurs.

Larvae with the queen genome but feeding on an infavourable food quantity develop into workers. Kerr and Nielsen (1966) reported that for *M. marginata* the queen genome of such workers is still expressed in the nerve cord: the last abdominal ganglia of queens are fused while in workers they are separated. Of a total of 47 bees 35 worker bees had five abdominal ganglia, while 6 workers and 6 queens had four ganglia. In other species, however, both queens and workers may have 4 or 5 ganglia. Apparently the relation between the queen determining genes and the fusion of the ganglia is lost in these species (Kerr 1974, Darchen and Delage-Darchen 1975).

Interesting results were obtained by Lemasson Naves da Silva (1973). She exchanged the queens from colonies of the two subspecies *M. quadrifasciata quadrifasciata* and *M. q. anthidioides*. The latter is a more northern subspecies with a slightly greater body size. In one out of four such queen transfers the colony with the *anthidioides* workers and *quadrifasciata* queen produced about 50 % queens (Table 6). The table also shows the frequent occurrence of 25 % queens.

TABLE 6 The effect of queen exchange between colonies of *Melipona quadrifasciata quadrifasciata* and *M. q. anthidioides* on caste of female offspring. After da Silva 1973.

	before exchange of queens		after exchange of queens	
workers of subspecies	number of females	percentage queens	number of females	percentage queens
M.q.q.	53	26	200	15
M.q.a.	25	28	199	48
M.q.q.	100	27	100	20
M.q.a.	100	22	100	23
M.q.q.	100	16	100	22
M.q.a.	100	18	100	19
M.q.q.	200	7.5	200	8
M.q.a.	200	10	200	16

In a recent paper Camargo, Almeida, Parra and Kerr (1976) report on *in vitro* rearing of four species, *M. quadrifasciata, marginata, rufiventris* and *scutellaris*. In those series where more than the natural mean quantity of food was given the percentages of queens rose to about 25 % (Table 7) thus supporting the theory of Kerr. Similar data on *quadrifasciata* have been reported by Velthuis and Velthuis-Kluppell (1975) who found 6 queens and 36 workers by *in vitro* rearing with 136 % of the natural food quantity in their control series.

TABLE 7 Queen-worker ratios in four species of *Melipona* under supernormal food conditions. After Camargo *et al.* 1975.

species	food quantity (% of natural mean)	number of queens	number of workers
M. quadrifasciata	125	9	41
quadrifasciata	175	3	16
rufiventris	125 - 150	12	29
marginata	120 - 130	9	19
scutellaris	above 100 %	5	13
Total for all species		38	118

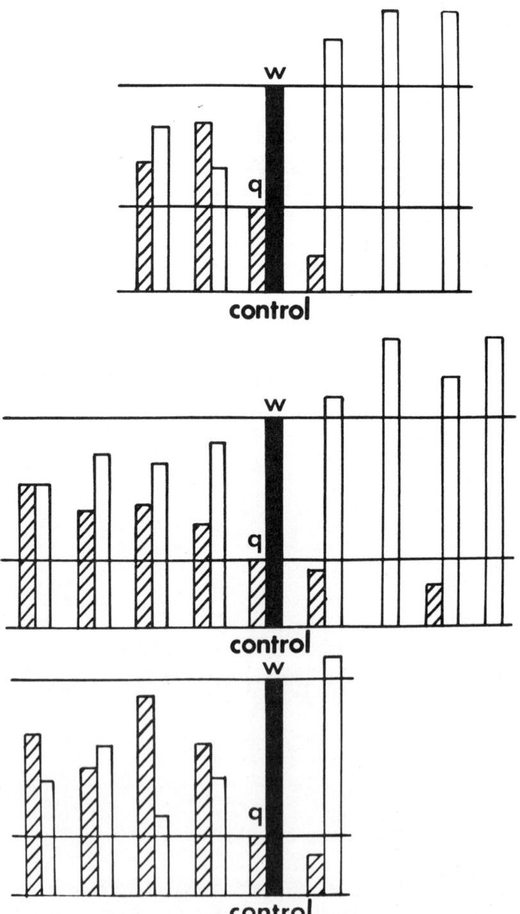

Fig. 3. The percentages of queens occurring in three experiments of Darchen and Delage-Darchen (1975). In the central row the percentage queens (workers in black) that has been found in untreated animals is given; at the left hand side the effect of extra food, at the right hand side that of reduction of food is given.

In complete disagreement with the above findings is the conclusion reached by Darchen and Delage-Darchen (1974, 1975). They studied the Mexican species *M. beecheii* and reduced or increased the natural food quantity either by taking the larvae from their cells before food uptake was completed or by injection of additional food in the cell. Although these procedures do not permit any quantitative description of the interference and although data such as pupal weight are not presented, there appears to be a good correlation between the amount of food and the percentage of queens in the offspring. Up to 71.4 % of queens is obtained. Their data are summarized in Fig. 3.

This paper of the Darchens can be critisized on various points. I will raise three objections. First, their experimental and control animals are from different brood batches and have a different production period. It has long been known that the percentage of queens fluctuates and shows a distinct seasonality. Therefore the control data (all about 25 % queens) are no good basis for comparison. Furthermore, their own statistical analysis seems to be an improper use of the χ^2 test. Even more serious is that they give no numerical values but only percentages. This prevents any alternative interpretation and also a statistical comparison with the Kerr hypothesis. Because of these shortcomings the paper must be regarded as being fully inadequate to disprove this hypothesis.

ENDOCRINE ASPECTS OF CASTE DETERMINATION

Little is known of the endocrine regulation of caste determination. As in the honeybees, the stingless bees show differences in corpora allata size. For the group as a whole, Akahira, Beig and Kerr (1967) studied biometric and histological properties of the corpora allata in adult workers for 32 species. They notice an inverse correlation between activity of the corpora allata and the capacity of the species to produce worker eggs. Beig and Baldissera (1974) continued such studies.

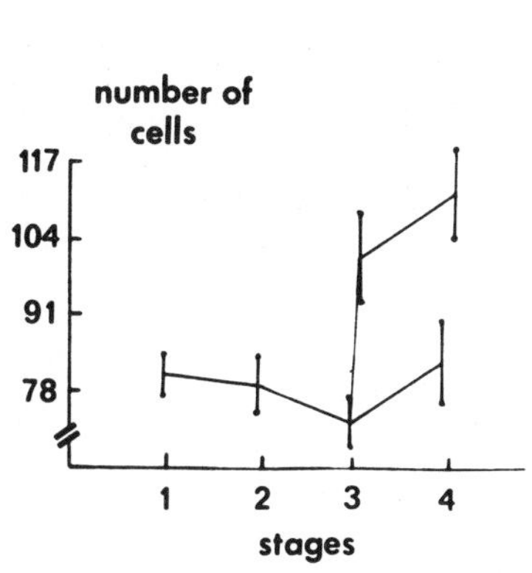

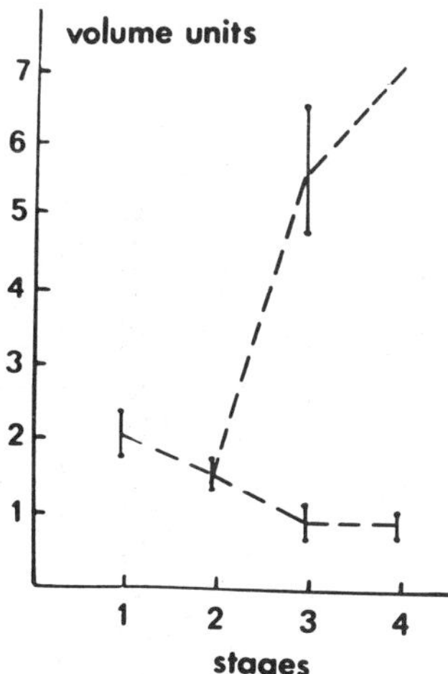

Fig. 4. The number of cells and the volume of the corpora allata in relation to development of *M. quadrifasciata*. After Kerr, Akahira and Camargo (1975).
1: prepupa just after the spinning of the cocoon but before defaecation.
2: prepupa in which the legs are almost developed but still folded.
3: pupal stage with black eyes and with the beginning of body pigmentation.
4: recently emerged adults.

Kerr, Akahira and Camargo (1975) measured volume and number of cells in the corpora allata in relation to pupal development. Fig. 4 shows their results: the larvae that just completed spinning the cocoon all have about the same corpora allata volume, but when the castes can be distinguished by the development of the eye, distinct differences show up. Apparently, queen development is correlated with greater corpora allata activity as in the other bees studied. The authors try to separate the values for volume into a bimodal distribution, reflecting differentiation into workers and queens. However tempting this may be, the result is not quite convincing (Fig. 5), although such a development must take place on the basis of Fig. 4.

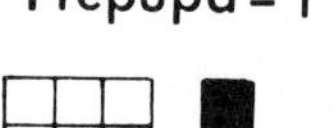

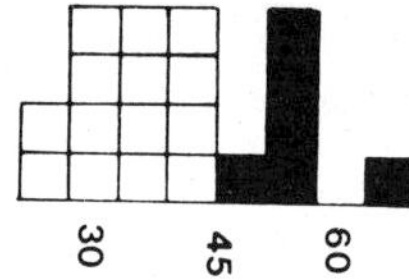

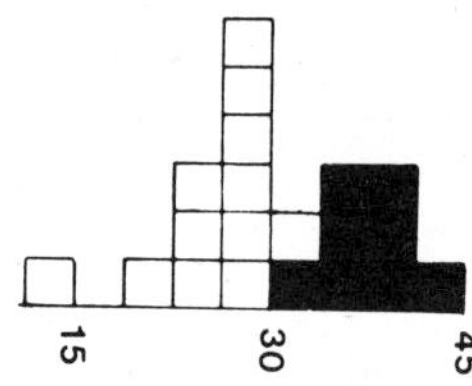

Fig. 5. The 3:1 segregation in the prepupal stages (1 and 2 of the preceding Fig.) according to Kerr, Akahira and Camargo (1975). In the prepupa 1 the authors suggest that 14 workers and 6 queens will develop; for prepupa 2 this should be 13 and 8 resp. In these stages no other characteristics permit a distinction between worker and queen.

Campos, Velthuis-Kluppell and Velthuis (1975) and Velthuis and Velthuis - Kluppell (1975) studied the effect of Juvenile Hormone (JH) application on larvae and prepupae of *Melipona quadrifasciata*. They found the early prepupal stage to be a very sensitive moment. Of the several components tested for JH activity the Zoecon products ZR 512 and ZR 515 were effective in the range of 10 mmg/larva, while the isomer mixture of Hoffman-LaRoche even at a dose of 10^{-7} mmg/larva caused larvae to differentiate into queens. Up to 100 % of the treated larvae became queens.

Interesting is the finding that no intermediate adults were found (Fig. 6 and 7), but some animals lost much body weight and became miniature workers. Miniaturisation was much more important in workers than in queens.

The result of one experiment of Velthuis and Velthuis-Kluppell (1975) demonstrates the effect of an unknown environmental factor (Table 8). With the lowest concentration 50 % queens were produced. The experimental animals came from different colonies and were grafted on different days. Furthermore, the speed of their development differed. The animals grafted on 24.4 have a relatively slow larval development. When treated subsequently with a low dose of JH, only one became a queen. A treatment with a high dose of JH, however, produced only queens although these animals have the same growth retardation. It seems as if growth rate is inversely correlated with JH sensitivity. A genetic explanation is not needed here.

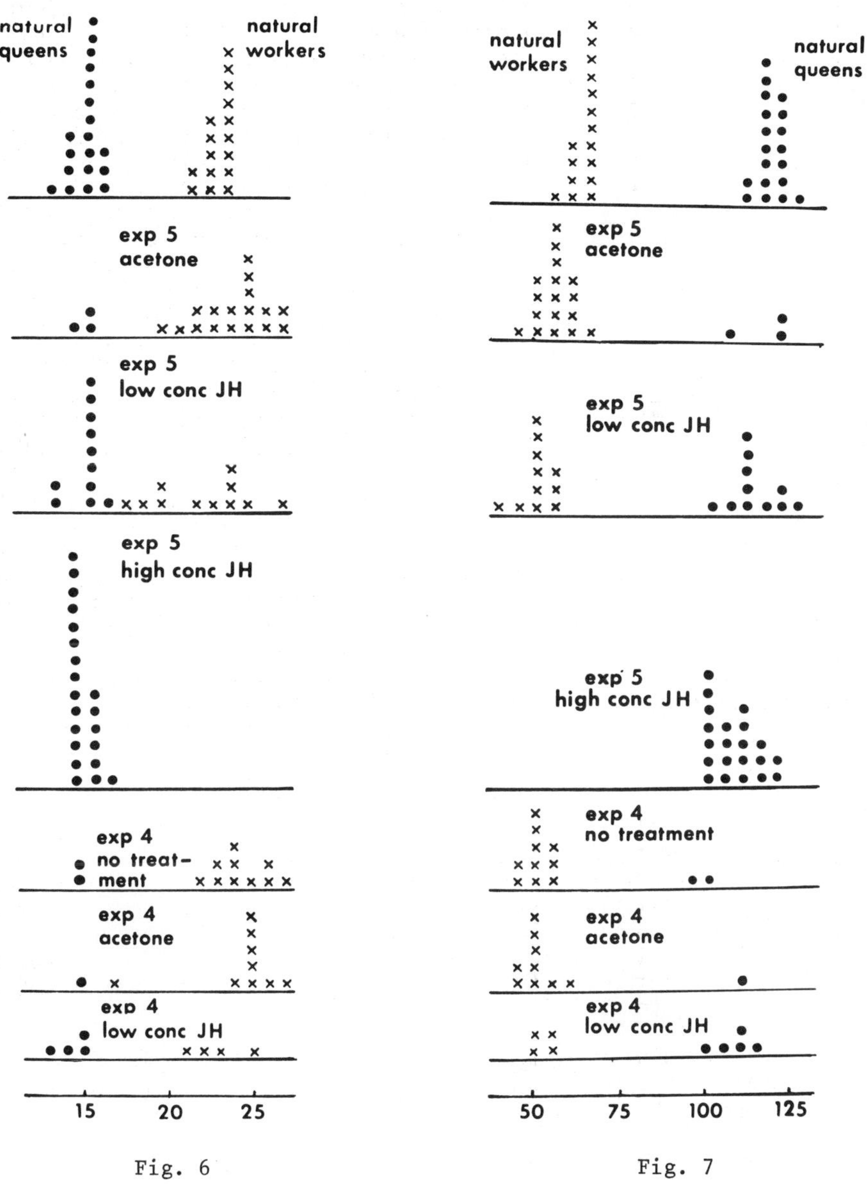

Fig. 6 Fig. 7

Fig. 6. The maximal width of the hind tibia in *M. quadrifasciata* of the animals obtained with JH application by Velthuis and Velthuis-Kluppell (1975). Some miniature workers approach the size of the queens.

Fig. 7. The distance between the margin of the eye and the mandible in *M. quadrifasciata* under the influence of JH application. From Velthuis and Velthuis-Kluppell (1975).

TABLE 8 The relation between the duration of larval development and the effect of JH application on caste development in *Melipona quadrifasciata*. Unpublished data of Velthuis and Velthuis-Kluppell.

treatment of early prepupae	caste of emerging adults	
	queen	worker
2.10^{-3} mmg HJ	22	0
2.10^{-7} mmg HJ	11	11
acetone	3	17

	duration of larval development				becoming	
	4	5	6	7 days	queen	worker
high concentration						
grafted on 24.4.74		2	6	1	9	
25.4		7	1		8	
26.4	1	1	3		5	
becoming queen	1	10	10	1	22	
low concentration						
grafted on 24.4.	1		6	1	1	7
25.4		8			6	2
26.4	2	3	1		4	2
becoming queen	2	8	1		11	
worker	1	3	6	1		11
acetone						
grafted on 24.4		1	5	2		8
25.4		7			2	5
26.4	2	2	1		1	4
becoming queen		3			3	
worker	2	7	6	2		17

DISCUSSION

Food as the determining factor in Meliponids

In both the Trigonini and the Meliponini the food deposited into a cell is produced by a number of workers. This number may be small, like 2 in the primitive *Hypotrigona muelleri* (Sakagami and Zucchi 1974), or in the order of 5 - 15 like in most other species. In the Trigonini the quantitative difference in the food for worker or queen is in the order of magnitude of 2 or more, in the Meliponini this variation is far less. The quantitative factor therefore could play a much more decisive part in the Trigonini than in the Meliponini. Nevertheless the possibility of qualitative differences should be considered.

The food available for a larva after it hatches consists of a liquid layer, the glandular material mixed with honey, and a more solid part where sedimented pollen grains are mixed with the same liquid. Since the freshly hatched larva floats on the food, the liquid layer is consumed before the developing larva reaches the more solid material. Therefore, any quantitative difference might be related to qualitative differences in the emerging adults, since with more food the larva spends more time on the liquid layer.

The emerging queens from colonies of *Scaptotrigona postica* are on the average 2.27 times the weight of emerging workers. Multiplication of the food quantity for a worker, 35.4 mg, with this factor gives 80.4 mg. Table 3 shows that with such amounts most larvae develop into queens. When we suppose that in *Scaptotrigona*, like in *Apis* and *Melipona*, the conversion of larval food into adult body weight is less efficient in queens than in workers, it appears that in the natural situation the quantitative factor alone already suffices to explain the differential development of the two castes.

The regulation of female dimorphism in *Melipona* must depend upon another mechanism and therefore, next to the genetic mechanism, the probability of a qualitative difference in the food must be considered. Indeed the Darchens (1975) have developed a model on that basis. They suppose that differences exist in the concentration of a factor probably originating from the food glands. Apparently, in that case some workers add more of this factor to the cell than others. Since several workers are needed to provision a cell, we might expect all kinds of concentrations of such a factor instead of only two types of food, with or without it. Unless the discrimination mechanism is very sensitive to the critical concentration we expect a lot of intercastes or intermediates to occur naturally. This is not the case. Furthermore, the differentiation would start early, like in *Apis*. This also is not the case.

When, however, a differentiation mechanism is put to work by a quantity of food, the decision will fall late in larval life, depending on whether or not the amount of food surpasses the critical level. To discriminate, again a sensitive mechanism has to be present. It is within this, probably narrow, region of food quantities that qualitative food factors may influence the way of further development.

Our results with topical application of JH show that the animals are most sensitive at the end of larval life. Differences in corpora allata size become apparent in the prepupal stage (Kerr *et al.* 1975). These findings point to a late moment of differentiation.

A critical question concerning a qualitative food factor in the absence of a quantitative one versus the genetic mechanism is the occurrence of intercastes. The Darchens (1975) report the appearance of such animals in their experiments. At the same time they remark that, however distinct they may be, they always closely approach one of the extremes, queen or worker. It is just such a close resemblance with the natural animals that led Velthuis and Velthuis-Kluppell (1975) to conclude the absence of intercastes. With miniaturisation the reduction of several body parts will not be of the same degree.

The genetic model of caste and sex determination

Originally Kerr (1950) supposed the presence of 3, later of 2 genes each with two alleles to be responsible for queen differentiation, where only the heterozygotic form would have the capacity to develop into a queen. The present interpretation of this concept and its further elaboration is described in Kerr (1975a and 1975b) and Kerr, Akahira and Camargo (1975), the latter being the ultimate formulation (Fig. 8). Now the caste determining genes are included in the sex determining ones, which have already been known for the honeybee for a longer time. The model is based on the papers of Britten and Davidson (1969, 1971). Three batteries of genes are proposed. The first consists of the two caste determining genes X^a and X^b responsible for JH production. The coding for RNA of each of the alleles of these two genes would be incomplete, and therefore only the double-heterozygote is able to synthesize the complete RNA. Probably we may interprete the alleles for X^a and X^b to be degenerations of the original gene that should be present in the intact form in *Apis*. They form part of the genes for femaleness.

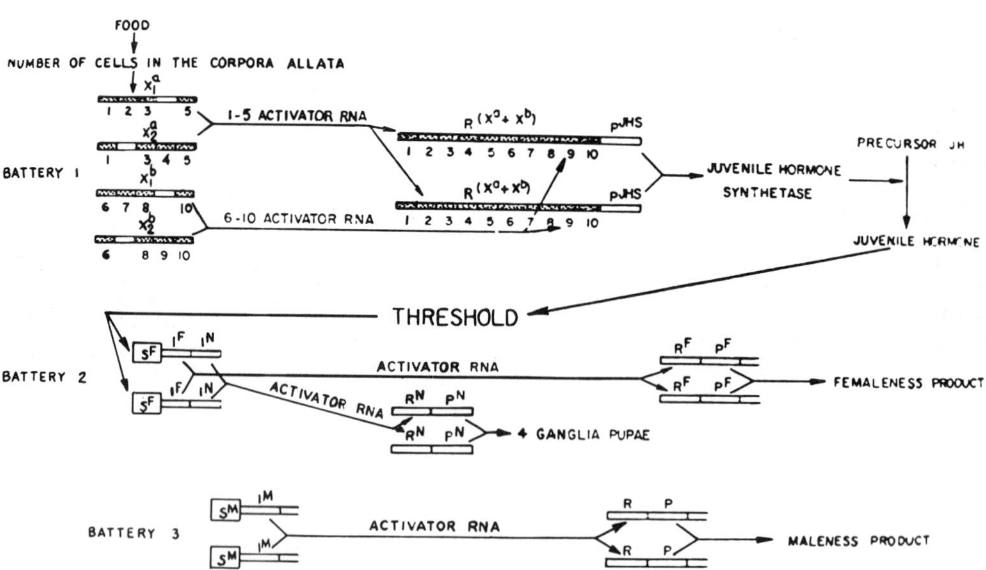

Fig. 8. Scheme of the genetic model for caste and sex determination in bees. From Kerr, Akahira and Camargo (1975).

Battery two consists of the other genes for femaleness. In early embryonic development the genes of battery two determine the sex of the developing larvae. After the uptake of all the larval food the food quantity determines the activity of battery one, which leads to the production in high or low quantities of JH by the corpora allata. Only the high JH titres re-activate battery two, who then determine the caste to be queenlike. When, however, this reactivation fails to occur, the genes of battery three (genes for maleness) express their influence in the male-like external morphology of the resultant adult worker.

When, because of inbreeding, the sex alleles of battery two become identical, the diploid individual becomes a male, since then battery two is unable to function. Diploid drones have first been described by Woyke (1963, 1965) for the honeybee and are later on also described for the Brazilian bumblebee *Bombus atratus* by Garófalo (1973) and by Camargo (1976) for *Melipona quadrifasciata*.

No doubt this daring and imaginative theory has to be provided with a lot more experimental evidence. Since part of the theory is purely genetic, the most decisive proof has to be presented by genetic experiments concerning the mendelian inheritance of the X^a and X^b alleles and their effects. However, the theory also involves developmental and endocrine processes and also in this field a lot of new experiments will be initiated by the theory.

A detail to be explained is the presence of a 3:1 segregation in *M. quinquefasciata*. This animal is peculiar in having n = 18 chromosomes while in other *Melipona* species studied n = 9 has been found. This polyploidy could be advantageous for the caste determination for it would reduce prospective queen numbers to 1:15 if 4 genes govern this. It would reduce the waste of the production of too many surplus queens. Kerr (1974) interpretes the 3:1 segregation of *M. quinquefasciata* as an indication that the genetic control mechanism in *Melipona* arose rather late in evolution, namely only after the polyploidisation and therefore independently in different species. The question that arises then is how this mechanism in these other species evolved. Are here also independent novelties? Why did it always turn out to be a set of two genes each with always two alleles, and why did the mechanism not arise in any other bee? To me the genetic mechanism seems so closely related to the identical food quantities of worker and queen larvae that the mechanism should arise before the diversification of the genus.

Now that by application of JH all female genomes for X^a and X^b can be turned into queens and the method of controlled mating for *M. quadrifasciata* is known (Camargo 1972c), the tools for direct genetic proof are available. Queens with 0, 25, 50 or 100 % potential queens in their offspring can be expected. Such a result will fortify strongly the general acceptation of the Kerr theory.

REFERENCES

Akahira, Y., D. Beig and W.E. Kerr, Corpora allata in Brazilian stingless Bees, J. Hokkaido Univ. Educ. (IIB) 18, 24-42 (1967).

Beig, D., The production of males in queenright colonies of *Trigona (Scaptotrigona) postica*, J. Apicult. Res. 11, 33-39 (1972).

Beig, D. and S. Baldissera, Controle endócrino nos meliponíneos. I. Atividades dos corpora allata e desenvolvimento dos ovários de *Melipona quadrifasciata* Lep. (Hym. Apidae), Ciência e Cultura 26, 1155-1160 (1974).

Britten, R.J. and E.H. Davidson, Gene regulation for higher cells: a theory, Science 165, 349-357 (1969).

Britten, R.J. and E.H. Davidson, Repetitive and non-repetitive DNA sequences and a speculation on the origins of evolutionary novelty, Quart. Rev. Biol. 46, 111-133 (1971).

Camargo, C.A. de (1972a) Aspectos da reprodução dos Apideos sociais. Masters thesis, Ribeirão Preto, Brazil.

Camargo, C.A. de (1972b) Produção 'in vitro' de intercastes em *Scaptotrigona postica* Latreille. In: Homenagem à Warwick E. Kerr, p. 37-54. Rio Claro, Brazil.

Camargo, C.A. de, Mating of the social bee *Melipona quadrifasciata* under controlled conditions, J. Kansas Entomol. Soc. 45, 520-523 (1972c).

Camargo, C.A. de (1976) Doctors thesis, Ribeirão Preto, Brazil.

Camargo, C.A. de, M.G. de Almeida, M.G.N. Parra and W.E. Kerr (1976) Genetics of sex determination in bees IX Rate of queens and workers produced by larvae raised under controlled conditions, J. Kansas Entomol. Soc., in press.

Campos, L.A. de O., F.M. Velthuis-Kluppell and H.H.W. Velthuis, Juvenile hormone and caste determination in a stingless bee, Naturwiss. 62, 98-99 (1975).

Darchen, R. and B. Delage, Facteur déterminant les castes chez les Trigones (Hyménoptères Apidés), C. R. Acad. Sci. Paris 270, 1372-1373 (1970).

Darchen, R. and B. Delage-Darchen, Le déterminisme des castes chez les Trigones (Hyménoptères, Apidés). Insectes Sociaux 18, 121-134 (1971).

Darchen, R. and B. Delage-Darchen, Nouvelles expériences concernant le déterminisme des castes chez les Mélipones (Hyménoptères, Apidés), C. R. Acad. Sci. Paris 278, 907-910 (1974).

Darchen, R. and B. Delage-Darchen, Contribution à l'étude d'une abeille du Mexique *Melipona Beecheii* B. (Hymenoptère: Apide), Apidologie 6, 295-339 (1975).

Garófalo, C.A., Occurrence of diploid drones in a neotropical bumblebee, Experientia 29, 726-727 (1973).

Kerr, W.E., Formação das castas no gênero *Melipona*. An. Escola Superior Agr. 'Luiz de Queiroz' 3, 299-312 (1946).

Kerr, W.E., Genetic determination of castes in the genus *Melipona*. Genetics 35, 143-152 (1950).

Kerr, W.E. (1974) Geschlechts- und Kastendetermination bei stachellosen Bienen. In: Sozialpolymorphismus bei Insekten, G.H. Schmidt, Ed., Wiss. Verlagsges., Stuttgart. p. 336-349.

Kerr, W.E., Sex determination in bees III Caste determination and genetic control in *Melipona*, Insectes Sociaux 21, 357-368 (1975a).

Kerr, W.E., Evolution of the population structure in bees. Genetics 79, 73-84 (1975b).

Kerr, W.E., Y. Akahira and C.A. de Camargo, Sex determination in bees IV Genetic control of juvenile hormone production in *Melipona quadrifasciata* (Apidae), Genetics 81, 749-756 (1975).

Kerr, W.E. and R.A. Nielsen, Evidences that genetically determined *Melipona* queens can become workers, Genetics 54, 859-865 (1966).

Kerr, W.E., A.C. Stort and M.J. Montenegro, Importância de alguns fatôres ambientais na determinação das castas do gênero *Melipona*. An. Acad. Bras. Ciencias 38, 149-168 (1966).

Michener, C.D. (1974) The social behaviour of the bees, Harvard Univ. Press, Cambridge Mass., 404 p.

Nogueira-Neto, P. (1970) A criação de abelhas indígenas sem ferrão, Chácaras e Quintais, São Paulo.

Sakagami, S.F. and R. Zucchi, Oviposition process in a stingless bee, *Trigona (Scaptotrigona) postica* Latreille, Studia Entomol. Rio de Jan. 6, 497-510 (1963).

Sakagami, S.F., C. Camilo and R. Zucchi, Oviposition behavior of a Brazilian stingless bee, *Plebeia (Friesella) schrottkyi*, with some remarks on the behavioral evolution in stingless bees, J. Fac. Sci. Hokkaido Univ. VI 19, 163-189 (1973).

Sakagami, S.F. and R. Zucchi, Oviposition behavior of two dwarf stingless bees, *Hypotrigona (Leurotrigona) muelleri* and *H. (Trigonisca) duckei*, with notes on the temporal articulation of oviposition process in stingless bees. J. Fac. Sci. Hokkaido Univ. VI 19, 361-421 (1974).

Silva, D. Lemasson Naves da (1973) Estudos bionômicos em colônias mistas de Meliponinae (Hymenoptera, Apoidea), Doctors thesis, Ribeirão Preto, Brazil.

Terrada, Y. (1973) Contribuição ao estudo da regulação social em *Leurotrigona muelleri* e *Frieseomelitta varia* (Hymenoptera, Apidae), Masters thesis, Ribeirão Preto, Brazil.

Velthuis, H.H.W. and F.M. Velthuis-Kluppell, Caste differentiation in a stingless bee, *Melipona quadrifasciata* Lep., influenced by juvenile hormone application, Proc. Kon. Ned. Akad. Wetenschappen Ser. C 78, 81-94 (1975).

Wagner, H.O., Haustiere im vorkolumbischen Mexico, Z. Tierpsychol. 17, 363-357 (1960).

JUVENILE HORMONE AND QUEEN REARING[+] IN BUMBLEBEES

Peter-Frank Röseler
Zoologisches Institut, Lehrstuhl für Zoologie II und vergleichende Physiologie, Röntgenring 10, 8700 Würzburg, Germany

The structure of castes

In the bumblebee species *Bombus hypnorum* and *Bombus terrestris* the castes differ mainly in physiology (Ref 1, 2). Body size is the only morphological difference between workers and queens. In *B. terrestris* there is a clear distinction in size between the castes: queens are always larger than workers. In *B. hypnorum*, in contrast, there are intermediate sizes between queens and workers, so that a worker can be larger than a small queen. From size alone the intermediates cannot be classified as either workers or queens. But in both species the castes are each different in physiology: Only young queens are able to overwinter and to start a new colony as a solitary bee during the next spring. In the very first day of adult life queens begin to synthesize reserve material in order to survive the winter. During the following days their fat bodies become very voluminous. In hemolymph the protein pattern differs from that of workers by a special fraction (Fig. 1) and lipid substances accumulate in the first two days to a level which is twice as high as that of workers. In addition, there are behavioural differences between the castes related to the different metabolism: on the second day of adult life bumblebee workers start to feed larvae, to secrete wax, to construct nest, and they are just able to collect food; moreover, they defend their colony. Young queens, in contrast, do not participate in these tasks, they do not even defend the nest but try to find refuges. After the fourth day most of queens leave their nest for the first time, but without collecting food. They are attractive for drones and ready to mate.

In this manner bumblebee queens and workers differ distinctly in physiology and behaviour. Therefore both are true castes. Experiments have revealed that the castes are alternative modifications of the same genome: a female larva is bipotent, it is able to develop into either a worker or a queen. Intercastes do not exist in either species.

[+] Supported by Deutsche Forschungsgemeinschaft.

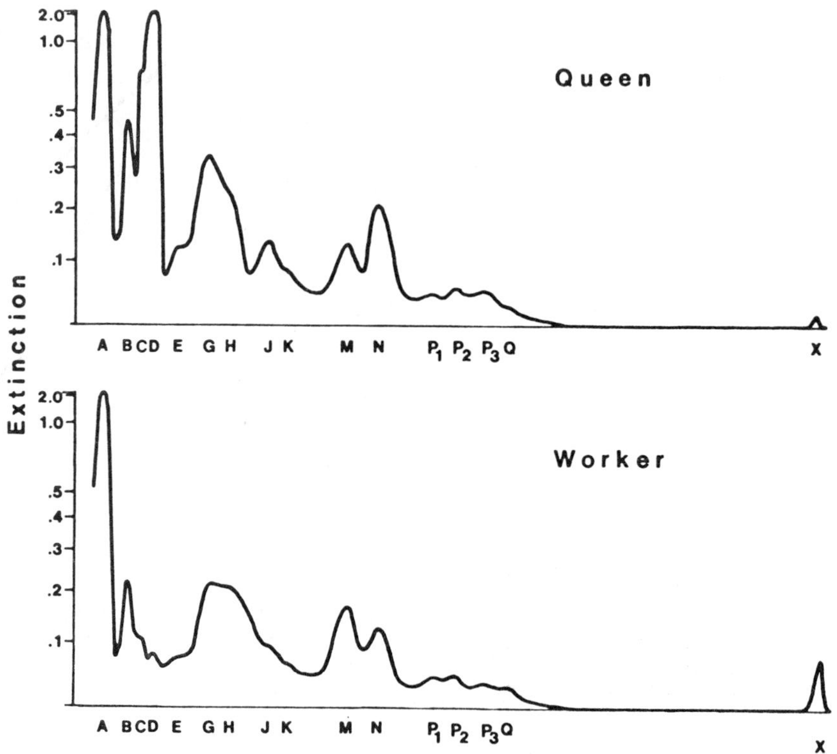

Fig. 1. Disc-electropherograms of hemolymph proteins of 2-day-old bumblebee queens and workers. The fraction "D" is specific to queens.

Such intercastes are reported of the brazilian species B. atratus by Camargo et al. (3), but a definition of the caste structure of that perennial species has not been given up to now.

The determination of castes

In B. hypnorum the divergent development of larvae is caused only by the amount of food larvae obtain (Ref. 1, 2, 4). Larvae will develop into queens if they are abundantly fed, and for this there must be an optimal worker - larvae ratio greater than 1 : 2. Observations of feeding frequency have shown that larvae of presumptive queens are more frequently fed than larvae of presumptive workers during the final instar. Up to now, there is no indication that in B. hypnorum any qualitative food factor exists responsible for caste determination. For example, none of the workers feed exclusivly either queen or worker larvae but often they change from one kind of larvae to the other during the same feeding visit. Moreover, the nurse bees of different age groups participate in feeding in accordance with their frequency within the colony.

In B. hypnorum development into a queen can still be induced by an abundant quantity of food during the final instar. In this case the larval development is prolonged. It is very likely, therefore, that a critical period exists at the end of final instar during which the development of larvae can be influenced. The amount of food itself cannot determine the development of a larva, this food signal must be transformed into a specific signal controlling the development; this signal must be produced by a regulating centre. Hence we suggested that the quantity of food above a certain limit activates the endocrine system of larvae resulting in a higher titre of that hormone which is responsible for the development of queens. This assumption was supported by the discovery that juvenile hormone titre is higher in presumptive queens than in presumptive workers at the end of final instar.

We have tried to induce queen development by topical application of synthetic juvenile hormones. The juvenile hormone mostly used was a C_{18} mixture of stereoisomers dissolved in acetone[+]. For each experiment the hormone was applied to some larvae situated at the periphery of a brood clump containing larvae of the same age. This treatment resulted in a decrease of hormone concentration from the treated larvae to the neighbouring larvae, which had probably obtained less hormone through the wall of cocoon, and finally to the distant larvae which hadn't got any hormone. The treatment had no effect when hormone doses up to 5 µg were applied only once. But quantities of 5 µg or 16 µg hormone given three times in the space of one day effected a delay of pupation; the larvae were fed further and they developed into large queens. This development was also induced in the neighbouring larvae, even when high doses of hormone caused pathological effects in the treated larvae. These pathological effects were as follows: cranial parts of the compound eyes were malformed and the wings were crippled; moreover, some of these larvae were not able to spin a cocoon in the regular manner and to adopt the upright stance during pupation.

In B. hypnorum juvenile hormone influences the development of larvae and can induce the development of queens. These experiments do not permit further conclusions, because we do not know whether the induction of queens is directly influenced by the hormone treatment or whether the elevated hormone titre only delayed pupation, thus permitting larvae to obtain more food and to develop into large queens. In addition, there is the possibility that the high juvenile hormone level does not result in queen development during the final instar but at a later period, for we had applied rather high doses, so that remains of the hormone could cause a higher level still later. In order

---

+ I am very thankful to Dr. R. Maag AG, Dielsdorf, Switzerland, for supplying various juvenile hormones.

to test this hypothesis we treated prepupae of presumptive workers. At this stage a single application of 0,1 µg - 1,0 µg juvenile hormone was sufficient to produce queens. These queens were of worker size but their physiology and behaviour were exactly queenlike, i.e. they did not participate in household tasks and they prepared for overwintering. Normally, queens of B. hypnorum have a wing-span of 34 - 40 mm. The smallest physiological queens obtained by the application of juvenile hormone to prepupae had a wing-span of 29 mm, a typical worker size. In smaller individuals there were no traceable effects.

In B. terrestris the larvae are determined as workers during the first instar by the presence of a dominant queen. After this determination the further larval development is immutably fixed and cannot be changed to queen development by great quantities of food. Topical application of juvenile hormone during the final instar delayed pupation up to 6 days, but all larvae developed into workers. Further comparative studies showed that even prepupae of presumptive workers did not develop into queens after hormone treatment. The effects brought about by juvenile hormone were malformations to a degree dependent on the hormone doses added: defects in the cranial parts of compound eyes, crippled wings, crippled legs and antennae, light spots in the chitin of corbiculae.

While in B. terrestris the development of larvae into workers is irreversibly fixed after the first instar, the development into queens can be experimentally changed up to pupation, resulting in bees which are workers physiologically, but are of queen size. This "redetermination" will take place when larvae of presumptive queens are introduced into a colony with a dominant queen. Because of this influence of the queen it is not possible to decide whether juvenile hormone is involved in determination during the first instar, since a larva possibly determined as a queen by the application of juvenile hormone will be redetermined by the dominant queen during its further development.

In the more primitive bumblebee species B. hypnorum larvae are determined as queens by the quantity of food only. Above a certain limit the amount of food seems to activate the endocrine system of larvae, resulting in an increased juvenile hormone titre which induces the determination as physiological queens during prepupal development. Normally, the characteristic size of queens is based upon this connection between the quantity of food and the determination as physiological queens. Only those larvae which receive much food become rather large and, on the other hand, they develop into physiological queens just by optimal nutritation. Therefore, the characteristic size of queens is an effect parallel to determination by food quantity.

This is also true for the more evoluted species B. terrestris, but larval development is additionally influenced by a dominant queen, which inactivates the endocrinous system of larvae probably by feeding them with a special qualitative factor; such larvae become workers.

In bumblebees, larval development into workers or queens seems to depend on a certain endocrinous program being alternatively switched by the quantity of food and in B. terrestris, in addition, by the quality of food. Later on, the determination itself regulates a specific endocrinous program governing the physiology of the adult castes. In queens, for example, it controlls the preparation for entering hibernation, which does not depend in either species on environmental factors, and also starts under constant light and temperature conditions in a climate room. The physiological pattern of adult queens during the first days is based upon a low level of juvenile hormone. If the titre is increased by application or injection of juvenile hormone, queens will not develop their fat bodies, the lipid content of hemolymph will reach only the same level as in workers (Fig. 2),

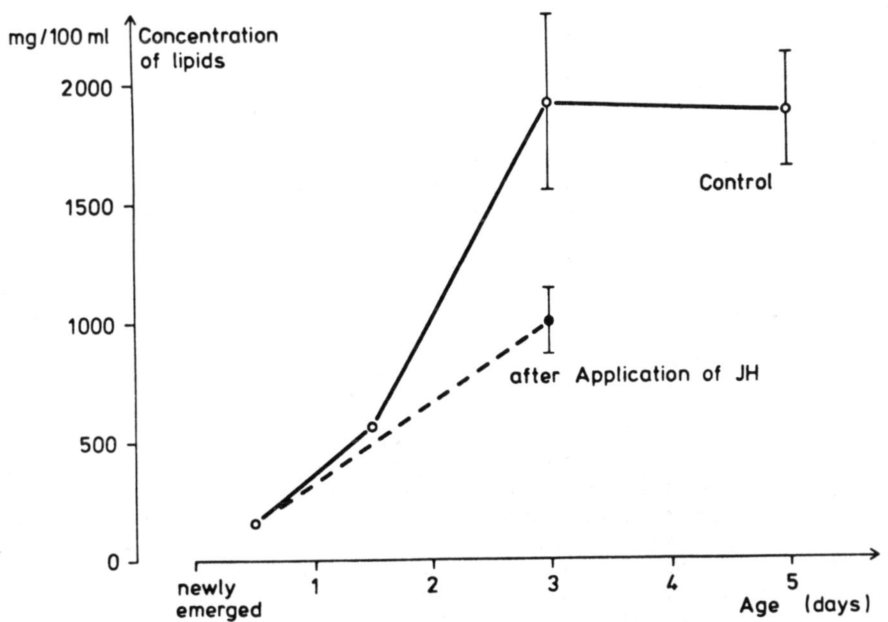

Fig. 2. Lipid concentration of hemolymph of bumblebee queens after injection of juvenile hormone.

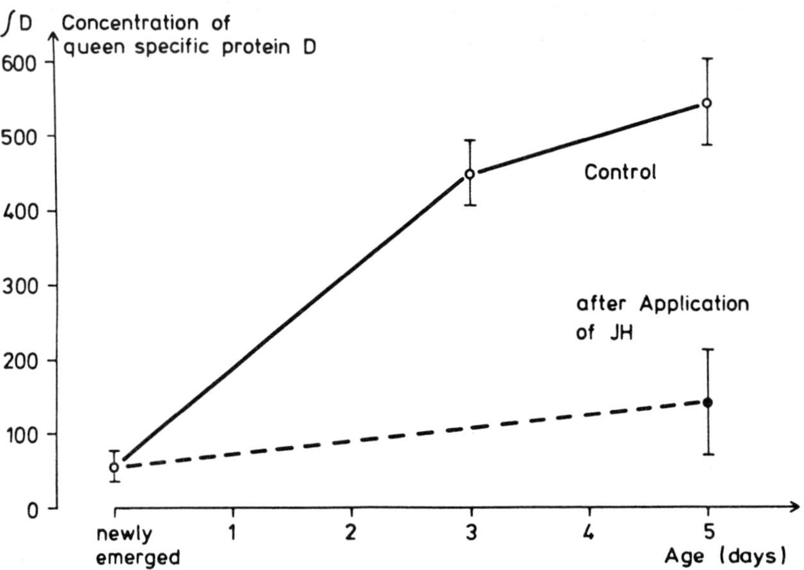

Fig. 3. Concentration of the queen-specific protein fraction "D" of hemolymph in queens after injection with juvenile hormone.

and the concentration of the queen-specific protein fraction will remain low (Fig. 3). Such queens treated with juvenile hormone do not leave their nest but participate in colony tasks just like workers. But unlike workers they are attractive for drones and ready to mate. Since oogenesis starts after a few days, such queens are physiologically similar to foundress queens in spring. This physiological changing of adult queens by hormone treatment during the first days can be obtained in both bumblebee species.

References

(1) Röseler, P.-F., u. I. Röseler, Morphologische und physiologische Differenzierung der Kasten bei den Hummelarten Bombus hypnorum (L.) und Bombus terrestris (L.), Zool. Jb. Physiol. 78, 175 (1974).

(2) Röseler, P.-F., Die Kasten der sozialen Bienen, Steiner-Verlag, Wiesbaden, 1975

(3) Camargo, C.A.de, Almeida, M.G.de, Parra, M.G.N. and W.E. Kerr, Genetics of sex determination in bees. IX. Frequencies of queens and workers from larvae under controlled conditions (Hym., Apoidea), J. Kans. Ent. Soc. 49, 120 (1976).
(4) Röseler, P.-F., Unterschiede in der Kastendetermination zwischen den Hummelarten Bombus hypnorum und Bombus terrestris, Z. Naturforsch. 25b, 543 (1970).

ENDOCRINE CONTROL OVER CASTE DIFFERENTIATION IN A MYRMICINE ANT

M. V. Brian
Institute of Terrestrial Ecology, Furzebrook Research Station, Wareham, England

INTRODUCTION

Little is known about the part endocrine organs play in ant polymorphism. In this contribution I propose to consider the ant *Myrmica rubra* in this connection, to discuss how the incidence of intercastes is reduced, and to relate its caste differentiation, as far as possible, to the morphogenesis of monomorphic insects.

MORPHOGENESIS

Post-embryonic Development

Eggs of *Myrmica rubra* L. are caste biased by a maternal influence that varies with the age of the mother, the rate of egg laying and the temperature. They hatch and pass through three larval stages before forming pharate pupae. Caste determination takes place in the third larval stage. Post-embryonic development in this stage has been charted in terms of larval fresh weight and the growth of those adult rudiments (imaginal discs) that can be seen through the transparent cuticle without damaging the larvae (1, 2). Leg and antennal buds, at first sub-spherical become ovoid and then segment transversely; wing buds grow from an ellipsoidal into a chordate shape. The cerebral ganglion (brain) migrates gradually from the cephalic capsule (head) into the thorax and becomes the focus around which the pupal head develops. There is not enough room in the larval head even for the antennal buds and it may be these which push the brain out. More likely perhaps, the transition is part of an organised developmental programme as stages in migration are clearly associated with other developmental events. The whole third larval stage divides into two major parts: one in which the brain slowly migrates and all the buds grow and one in which the legs and antennae grow length-wise and divide transversely. During the latter period the wings grow without dividing and the germarium of the ovary (though not visible) grows and divides into prospective ovarioles (in gynes not workers). These phases of development are referred to as the pre-segmental and the segmental periods respectively. The larvae next stops feeding, purges its gut and starts apolysis into a pharate pupa whose external parts burst their sheaths and lie in the ecdysial fluid between the old and new cuticles. The pre-segmental period is charted in terms of tenths of the brain migration; the segmental period in terms of the number of leg segments: one, two or three. These represent femur, tibia and tarsus; coxa and trochanter are formed from the sheath.

From the moment they enter the third stage the smallest larvae are certainly worker biased and perhaps worker determined for they never become gynes. The larger larvae grow and develop as far as brain position 5-tenths before another lot, again the smaller ones, become determined as workers and develop accordingly. More larvae change course in a society with queens than otherwise, for queens cause workers to feed these particular larvae abundantly with prey and water and so prevent diapause induction (3). The core of gyne biased larvae that remain grow very slowly indeed and may take several weeks to put on a few milligrams; nutrients are concentrated and stored in their fat body. Since their leg buds and brain change very little too, the gain per developmental stage is quite high, in the order of 0.5 mg, quite normal for gyne larvae at any stage. They stop in diapause when the brain is at 8-tenths of its transit. This lasts until winter when they slowly recover and by spring are able to resume very high rates of growth during which the adult rudiments are held in check so that gains of 0.75 mg/stage occur. The specific growth rate (growth per milligram of material) declines slowly throughout the spring until feeding stops but the specific growth rate of the wing buds is remarkably constant in gyne forming larvae until metamorphosis.

As spring growth is resumed the gyne biased larvae start emitting a chemical signal (4). If the colony is strongly queened this induces attack by workers and the resulting wounds combined with food dilution lead to worker morphogenesis (5, 6). The few larvae that survive or escape this treatment and reach the segmental period with gyne bias intact, stop signalling and are virtually certain of becoming gynes. This is a selective mechanism that is queen dependent; if there are no queens all the workers concentrate on rearing gyne biased larvae. By the time larvae have 2-segment legs but not before, the weight frequency distribution of those that form gynes is distinct from that of those that form workers; gyne larvae all exceed 5.4 mg at this stage. The same is true of wing area distributions. This population dissection is not very efficient; some 9% of intercastes occur. They clearly belong to the lower end of the gyne distribution not the upper end of the worker distribution; that is, they have passed the last worker gate but not managed to grow into gynes.

Worker Determination

The departure of workers from the gyne morphogenetic path leads to a premature neotenic metamorphosis. The first sign of worker determination is an accelerated leg growth or brain movement and an arrested wing and germarial growth. This happens quite suddenly and both enlarges the embryo relative to its larval host and gives it the chief structural features of the worker; no wings and one-tube ovaries. The new ratio embryo/total larva not only reduces assimilative power of the individual but ensures that the greater proportion of new nutrients is absorbed at once into embryonic growth instead of being stored in the fat body; usage instead of storage. A model of this has been considered by Skellam *et al* (7). The reduced values for gain/stage range from 0.2 to 0.3 mg compared with 0.5 to 0.8 mg for gynes. Notice that there is no overlap.

All worker determination takes place in the third instar; those with strong worker bias change early, either at the start or at 5-tenths brain; those which start with gyne bias, change at 8-tenths brain. At all stages it is the smaller larvae with relatively large legs and small wings, that is incipient worker shape, that change developmental course. Though all deviate in the pre-segmental stage the duration of the next, segmental stage is very variable: from three to over eight days; in complete contrast gynes and males regularly have a five day segmental period. The short three day period is especially a feature of diapause larvae that fail to recover or of those that are forced by high temperature culture (over 23°C) to metamorphose in autumn. The speedy segmentation compensates for the lateness of their conversion. The long periods are more a feature of very small larvae or larvae whose growth potential has been damaged by starvation.

ENDOCRINOLOGY

Histological studies have confirmed the typical function for the *pars intercerebralis* of the brain: it produces a material, staining purple with aldehyde fuchsin, which can be traced down axons to the *corpora paracardiaca*, where it is temporarily stored. It is released in worker biased larvae at about the stage when the brain is halfway (8) and in gyne biased ones later but not later than the end of the pre-segmental period (9). In the latter larvae this material is only produced in spring after recovery from diapause. It cannot be detected during segmentation in gyne forming larvae but it is still being produced in worker ones during this period. By comparison with other insects one can presume that it stimulates the production and dispersion of ecdysone though whether this is manufactured in the usual prothoracic glands or not is doubtful for though these are conspicuous in stage 2 larvae they cannot be found in stage 3 ones. Ecdysone may be manufactured in other parts of the body or stored from previous stages (10).

In *Myrmica* the role of juvenile hormone (JH) and moulting hormone (MH) has been tested using spring larvae fully recovered from diapause (11). Ejection of the brain through a slit in the dorsal thoracic wall stops feeding, growth and development. Ligature behind both the brain and *corpora allata* (CA) is followed by rapid leg growth and segmentation; the wing buds in contrast, stop altogether. If the JH analogues, farnesenic acid or farnesenyl-methyl-ether, are applied topically to such ligatured larvae, leg growth and segmentation is arrested. This is satisfactory evidence that JH inhibits the leg bud development programme as in other insects (10). As for the wings; either they are incompetent at this stage (and they certainly develop late, even later in females than in males) or, they need JH for growth. Such a differential effect on embryonic parts is quite a feature of JH control (12). The same JH analogues applied to normal growing larvae foster further growth and yield more, bigger gynes. From these results it is reasonable to conclude that worker determination is effected by stopping CA activity and destroying all JH.

The problem of what happens to gyne generating larvae remains. Presumably their CA do not switch off until much later. There is no sudden change in the relation of embryo to larvae during gyne genesis and it could well be that the CA just fade out gradually; they must have stopped secreting by the time the legs segment. By then, the wing and the germarium are probably competent to respond to metamorphosis hormones, and gyne form is determined.

Alpha-ecdysone has been tested on gyne larvae in spring. It has no perceptible effect on larvae that have not reached the segmental period, though it was suspected that their response threshold to later doses might have been lowered or the hormone stored (11). If injected during segmentation it accelerated the normal changes and caused earlier apolysis and metamorphosis. A higher proportion of workers and intercastes resulted and the gynes were smaller. This result resembles starvation of larvae before gyne determination but acts much more quickly. Thus ecdysone favours worker genesis and JH favours gyne genesis.

CASTE DIVERGENCE

It remains to consider what mechanisms are responsible for creating the marked discontinuity in caste size and structure. The main process is morphogenetic and will be considered first; others are social and concern population regulation.

Morphogenetic Regulation

The device of having two distinct growth/development ratios is not enough to ensure a discontinuity between workers and gynes; the period of worker determination must still be restricted and distinct from the period of gyne determination, otherwise all intermediate sizes and shapes giving intercastes will result. Obviously too, the bigger the interval the greater the distinction that can develop.

Such a zone undoubtedly exists. After the last worker gate at brain 8-tenths there are, for worker determined larvae three days of growth, whereas the gyne has six days: twice as many. If the gyne ratio of growth/development is double that of workers this enables four times as much gain in weight from the point on the gyne path from which they diverge. To what is this zone due? One plausible suggestion is that a moult has been suppressed so that whilst workers have three, queens have in effect four stages. This is supported by biometrical evidence based on the large body size relative to head capsule size that is attained only by the gyne forming larvae (11), and which of course accentuates the need for brain migration. Evidence points to a suppressed moult at a stage near 8-tenths brain; this is the size to which gyne biased larvae can creep in late summer before halting in diapause and it is also, of course, the stage of the last worker gate. In place of a moult there is cuticular expansibility, probably acquired overwinter at the same time as growth potentiality is restored. There thus appears to be reasonable evidence for a suppressed moult, that endows gyne-tending larvae with the additional time and internal capacity needed to establish a distinctive size and shape.

It is natural to go on to suggest that diversion of workers from the gyne path selects out the weaker individuals and leaves only the fittest, most vigorous ones for the reproductive role. There is ample evidence for this, since considering only the diapause larvae, it is the smaller ones with bigger leg buds and smaller wing buds that fail to sustain gyne growth in spring. Nevertheless a study of intercaste formation shows that the ability to avoid the worker gate does not guarantee development into a gyne.

There are three ways of forming intercastes artifically. All shapes between the worker and the gyne can be formed by culturing normally as far as the early segmental period and then starving completely. The longer the larvae grow beyond the start of this period the more gyne characteristics they assume on maturity. First, a worker exterior can contain increasing numbers of ovarioles; then, the wing can change from a stub to a tube and finally to the perfect blade. Starvation after the legs are 2-segmented simply produces small but perfect gynes (microgynes).

Intercastes can also be obtained by culturing diapause larvae in the process of recovery, that is midwinter; at that time they lack the power to sustain growth far enough to form full gynes. It is important to notice in this case that they are not diverted through the worker gate as would be expected if some selection mechanism were at work. This implies either that the worker determination mechanism needs to be established by diapause recovery at the same time as general growth ability is restored, or that it does not in fact filter the gyne potential larvae from the rest.

The third method of producing intercastes is to take post diapause larvae that are fully recovered and give them a period of incubation with all foods except protein. Only four days like this suffice to destroy all gyne potentiality; there is in fact a visible anisomeric change to a worker shaped embryo as already described. On incubation with full protein such larvae all yield worker pupae after a period of growth in the typical worker style with a ratio below 0.4 mg/stage. The actual growth rates vary considerably and segmentation may last between three and eight days but the essential characteristic is the fixed and small growth/development ratio which ensures that all individuals weigh the same when they become pupae. If instead of four days pre-treatment larvae are given two or three days, a high proportion of intercastes would be expected, if there was no gyne selection mechanism; or, a mixture of gynes and workers if there was. In fact a range from gynes through intercastes to workers results, and the frequency of intercastes is not significantly greater than the frequency obtained by normal rearing of untreated larvae (9%). What is greater than normal is the numbers that die; all in the segmentation period. There is thus no gyne selective mechanism operating. Many larvae without the ability to form gynes are clearly able to pass the worker gate, and go on to form either intercastes or die from internal confusion (13).

Population Regulation

From these results it follows that social and population processes must be involved in selecting individuals capable of sustaining growth on the gyne path once they have passed the last worker gate. This necessitates an ample supply of suitable food and attention at all times; larvae should also be totally quiescent (in diapause) when they cannot get this, at growth temperatures. Clearly regulation of the quantity of gynes produced is a pre-requisite for ensuring that they are all well fed. Queens exercise control over this in the three ways already mentioned. First some larvae acquire a worker bias from their mother, especially if she is young. Second, queens stimulate feeding and attention to small gyne biased larvae in the early third instar, so that many avoid diapause and after becoming worker determined at the stage brain 6-tenths, metamorphose. Third, queens suppress the majority of gyne potential larvae before they have passed the last worker gate, as after winter they emit a transient signal which enables workers to recognise them. With queens present, the workers then attack and bite them as well as diluting their protein food with water; as a result they metamorphose into workers. Any that escape this and reach the segmental period, when the signal stops (and some always do) attract ample food and grow well. Large larvae are intrinsically more attractive to workers than small ones and it is only the presence of queens that reverses this (14). By converting the majority of gyne tending larvae into workers the queens are acting as a valve which stops sexual production until the colony has the necessary trophic strength. Exactly how the queen control is blocked in opulent societies is as yet uncertain.

DISCUSSION

Worker determination has been shown to be a decline or cessation of CA secretion. This causes a growth of the embryo relative to its larval host, but a growth that is confined to a ventral set of imaginal discs that lie near the nerve cord; wings and ovaries, situated dorsally, are by contrast arrested. There is a well defined period in the last stage larva during which worker determination may occur, followed in those that escape, by four days of additional growth before gyne determination.

The discontinuity in caste size and structure in *Myrmica rubra* has several causes. There are two distinct morphogenetic styles, which diverge early enough to create a substantial difference. This divergence occurs in a well defined zone of development which may precede a suppressed moult. Finally population regulators vested in queens accentuate these divisory factors by creating a social pressure to become workers and so reducing the number of larvae that remain bipotent.

Gyne determination occurs at the 2-segment leg stage; after this starvation gives microgynes and before this intercastes. It seems to correspond morphogenetically with the 5 g weight of *Manduca sexta* (15, 16): after this starvation cannot prevent perfect pupae forming, below this down to 4 g larval/pupal intermediates mature. Also this is when the CA stops secreting JH in *Manduca*, and the larvae prepare to metamorphose rather than simply moult. The earlier, worker determination zone, is as would be expected, entirely lacking in *Manduca sexta*; one would have to look for a stage in which the CA stopped secreting and allowed neotenic metamorphosis to follow. In fact Nijhout and Williams produced evidence that brain competence to secrete prothoracotropic hormone (PTTH) only follows the termination of CA activity. This perhaps is odd in view of the fact that in earlier larval stages, PTTH is released and moulting follows though the CA does not stop secreting, or if it does, soon resumes again. In *Myrmica* PTTH is released from 5-tenths brain to the start of the segmental period.

As regards the triggering of metamorphosis as distinct from moulting: in *Manduca* the attainment of the weight 5 g appears to be the critical factor (15, 16); this value is calculated from the negative regression of pupal weight on time to develop. In *Myrmica*, the same relation has been demonstrated (17), and application of the same argument points to 4.2 mg as the critical metamorphosis weight for gyne forming larvae. This is the average weight of larvae at the start of their segmental period and the stage in development after which starvation cannot prevent pupation. However, as intercastes result from starvation between 4.2 and 5.4 mg (the average weight at the 2-segment leg stage) this weight is more equivalent to the 4 g of *Manduca*. An error may arise from the fact that in *Myrmica* the growth study started with hibernating larvae not with stage 2 larvae. The same growth rate/final weight argument applied to worker forming larvae indicates a diffuse zone, already known as that of divergence from the gyne development path, lying between 5-tenths brain and the end of the pre-segmental period.

So the two approaches do not quite give concordant results but the small discrepancies may eventually reveal interesting new problems for solution.

REFERENCES

(1) M. V. Brian, Regulation of sexual production in an ant society, Symposium No. 173 CNRS L'effet de groupe chez les animaux, Paris (1967).

(2) Brian, M. V. Kastendetermination bei Myrmica rubra L. Sozialpolymorphismus bei Insekten, (ed. G. H. Schmidt) Wissenschaftliche Verlagsgesellschaft, Stuttgart, 1974.

(3) M. V. Brian, Caste determination through a queen influence on diapause in larvae of the ant Myrmica rubra L., Ent. exp. and appl. 18, 429-442 (1975).

(4) M. V. Brian, Larval recognition by workers of the ant Myrmica, Anim. Behav. 23, 745-756 (1975).

(5) M. V. Brian, Feeding and growth in the ant Myrmica, J. Anim. Ecol. 42, 37-53 (1973).

(6) M. V. Brian, Caste control through worker attack in the ant Myrmica, Insectes soc. 20, 87-102 (1973).

(7) J. G. Skellam, M. V. Brian and J. R. Proctor, The simultaneous growth of interacting systems, Acta biotheor. 13, 131-144 (1959).

(8) J. S. Weir, Changes in the retro-cerebral endocrine system of larvae of Myrmica, and their relation to larval growth and development, Insectes soc. 6, 375-386 (1959).

(9) M. V. Brian, The neuro-secretory cells of the brain, the corpora cardiaca and the corpora allata during caste differentiation in an ant, Symp. "Ontogeny of Insects", Prague 167-171 (1959).

(10) Doane, W. W., Role of hormones in insect development. Developmental Systems: Insects (ed. S. J. Counce and C. H. Waddington) 2, 291-497 Academic Press, London and New York, 1973.

(11) M. V. Brian, Caste differentiation in Myrmica rubra: the role of hormones, J. insect physiol. 20, 1351-1365 (1974).

(12) J. H. Willis, Morphogenetic action of insect hormones, A. rev. Ent. 19, 97-166 (1974).

(13) M. V. Brian, Studies of caste differentiation in Myrmica rubra L. 4. Controlled larval nutrition, Insectes soc. 3, 369-394 (1956).

(14) M. V. Brian, Brood rearing behaviour in small cultures of the ant Myrmica rubra L. Anim. Behav. 22, 879-889 (1974).

(15) H. F. Nijhout and C. M. Williams, Control of moulting and metamorphosis with tobacco hornworm *Manduca sexta* L.: growth of the last instar larva and the decision to pupate, *J. exp. Biol.* 61, 481-491 (1974).

(16) H. F. Nijhout and C. M. Williams, Control of moulting and metamorphosis with tobacco hornworm *Manduca sexta* L.: cessation of juvenile hormone secretion as a trigger for pupation, *J. exp. Biol.* 61, 493-501 (1974).

(17) M. V. Brian, Studies of caste differentiation in *Myrmica rubra* 2. The growth of workers and intercastes, *Insectes soc.* 2, 1-34 (1955).

THE INFLUENCE OF JUVENILE HORMONE ANALOGUES ON CASTE DEVELOPMENT IN TERMITES

Ivan Hrdý
Czechoslovak Academy of Sciences, Institute of Entomology, Department of Insect Toxicology, Šalamounka 769, 150 00 Praha 5, Czechoslovakia

ABSTRACT

The juvenile hormone (JH) is responsible for cast formation in social insects. In termites, the development of the soldier caste, mainly of pre-soldiers, or soldier-worker intercastes, can be induced by treatment of pseudergate larvae by some of the synthetically prepared juvenile hormone analogues (JHA). The JH sensitive period depends on the intermoulting cycles. By contact or vapor treatment with JHA the soldier caste development can be induced at very early larval stages. In majority of individuals the soldier intercast development may be induced, when treated orphaned larvae - pseudergate groups in Petri dishes. In our tests, *Cryptotermes brevis*, *Neotermes castaneum*, *N. jouteli*, *Incisitermes schwarzi*, *Zootermopsis angusticollis*, *Z. nevadensis*, *Reticulitermes lucifugus*, *Prorhinotermes simplex*, *Heterotermes convexinotatus* and *Coptotermes formosanus* proved to be sensitive to JHA treatment (and several more species in other laboratories). In laboratory tests in presence of sexuals and soldiers, the induction of superfluous soldier formation by JHA treatment is much less successful and our preliminary field experiments with colonies of *Prorhinotermes simplex* showed no effect at given circumstances.

REFERENCES

(1) I. Hrdý, Der Einfluss von zwei Juvenilhormonanalogen auf die Differenzierung der Soldaten bei Reticulitermes lucifugus santonensis Feyt (Isopt.: Rhinotermitidae), *Z. angew. Ent.* 72, 129 - 134 (1972).

(2) I. Hrdý & J. Křeček, Development of Superfluous Soldiers Induced by Juvenile Hormone Analogues in the Termite, Reticulitermes lucifugus santonensis, *Insectes soc.* 19, 105 - 109 (1972).

(3) I. Hrdý, Effect of Juvenoids on Termites and Honeybees, Proc. VII Congr. IUSSI, London 1973, 158 - 160 (1973).

THE DEPENDENCE OF HORMONE EFFECTS IN TERMITE CASTE DETERMINATION ON EXTERNAL FACTORS

Michael Lenz*
Bundesanstalt für Materialprüfung, Unter den Eichen 87, D-1000 Berlin 45, Germany

1. INTRODUCTION

Several recent reviews inform about termite castes and the procedures of caste determination (Ref 23,26,31,34,35). The juvenile hormone (JH) has decisive significance in the caste regulation of termite colonies (Ref 17,18,23,26,47,48). The actual JH titre in a termite is not only its "own affair", and important only in the development of the individual, but is also influenced by occurrences in the colony, especially in its constitution and in the original caste composition, and becomes therefore dependent upon the requirements of the colony.

One way in which to show more lucidly parts of the many complex events exists in feeding JH analogues (JHA) to whole colonies, or groups of termites, or, as the case may be, in exposing them to the vapour phase of these substances. When the JHA concentration is increased, the demand for at least part of the caste regulating mechanism is strongly intensified. With suitable doses this effect also becomes visible.

From my own experimental data and from literature I report here on the effects of the environmental factor nutrition and of group composition during the procedures of caste formation, especially the differentiation of soldiers. Relatively few results on the significance of environmental influences and social structures are available (Ref 31).

2. EXPERIMENTAL APPROACH

Termite species used for the experiments were *Kalotermes flavicollis* (Fabr.), *Heterotermes indicola* (Wasmann), *Reticulitermes flavipes* (Kollar), *R. lucifugus* (Rossi), *Coptotermes amanii* (Sjöstedt), *C. niger* (Snyder) and *Nasutitermes nigriceps* (Haldemann). Only groups of definite composition from each species were set up, except for *K. flavicollis* where mainly whole colo-

*With support from the Deutsche Forschungsgemeinschaft (DFG)

nies were treated. All species had been held for years in the Berlin-Dahlem laboratory (Ref 4).

The two JHA's were ethyl-3,7,11-trimethyldodeca-2,4-dienoate (ZR-512, Altozar)** and isopropyl-11-methoxy-3,7,11-trimethyl-dodeca-2,4-dienoate (ZR-515, Altosid)**. The substances were mainly given on filter paper, but in some cases either on sound pine wood (Pinus) or such decayed by the brown rot fungus Poria monticola. N. nigriceps received JHA on sawdust from applewood (Malus). Further data on experimental procedures are given elsewhere (Ref 20,21).

3. INFLUENCE OF THE ENVIRONMENTAL FACTOR NUTRITION ON CASTE FORMATION

Among the factors which influence the vitality and growth rate of a termite colony, the quality of the available nutrition is of decisive importance. The caste proportions are adapted to the actual circumstances (Ref 2). For example, when the number of individuals in a colony increases, new soldiers are formed to restore the optimal and most "economical" ratio of soldiers to the other castes (Ref 43). With retarded growth or increased mortality rates, excess soldiers are actively eliminated or starved (Ref 7,9,32; see Chap. 4), so that over longer periods the respective relationships for each species remain relatively constant (Ref 13,27,34).

Specific food situations can clearly favour the development of individual castes. The xylophagous termites prefer wood infected and partially decayed by microorganisms. The Basidiomycetes of the brown rot type have proved to be especially advantageous (Ref 3,5,36). With exclusive feeding on such wood, the protein content of which is i.a. superior to that of undecayed wood, the number of late stage nymphs, alates, and neotenics, may increase (Ref 2,3). Particularly Heterotermes indicola reacts with an intensified formation of neotenics and an increased reproductivity rate when the insects receive wood decomposed through brown rot fungi (Ref 22). Deviations are shown to be determined largely by the species-specific reactions of the termites to the wood types, fungal species and strains, or the degree of decay in the wood utilized.

In laboratory groups the extent of soldier formation in Reticulitermes flavipes (Ref 8) could be connected to the quality of the food, i.e. with or without decay, and in Coptotermes formosanus (Ref 38) to the composition of its symbiotic protozoa fauna (and so finally also to the nutrition).

A deterioration of the nutritional situation in groups of Reticulitermes santonensis with functional reproductives, suppresses or retards the production of further sexual stages and also

**I wish to thank Zoecon Corp., Palo Alto, California, for the donation of these preparations.

that of soldiers (Ref 7). The more pseudergates colonies of Kalotermes flavicollis contain, the better the pair is provided with food and the greater is their stimulating effect on the formation of soldiers (Ref 43,44). The quality of food is, of course, likewise important.

In this context, it is also of interest that Nasutitermes exitiosus, which were either field-collected or held for some time in the laboratory, show significant differences in body measurements. This may also be attributable to differing nutrition (Ref 29).

Independently from the season, laboratory cultures of Kalotermes flavicollis, kept in Berlin-Dahlem, form alates when most of the wood reserves are eaten and not timely replenished.This could also be one way in which the number of individuals of a colony adapts to the available food supply.

The investigation of the effects of JHA on termites also revealed (Ref 20), that the insects were able to compensate these effects more effectively when the quality of the food was improved. Kalotermes colonies which received JHA-treated wood, with or without fungal decay, showed finally, despite doubled JHA consumption, no difference in the numbers of newly formed soldiers compared with colonies kept on sound wood. Decayed wood additionally reduced the total mortality rate of the colony. The difference in the constitution of both groups was significantly maintained even one year after the experiments were discontinued and both groups were equally well fed.

Similar observations were made on Coptotermes amanii. A further "compensatory mechanism" appeared with treated fungal wood where, despite a greater JHA intake, only intercastes worker/soldier were predominantly formed instead of soldiers, as was in the case of undecayed wood with JHA additives.

Figure 1 shows a further example of the effects of the quality of the available nutrition on processes in K. flavicollis colonies. The mortality rate of existent soldiers in the colonies was clearly related to the formation of new soldiers after feeding on Altozar, when the nutritional conditions were gradually improved (filter paper, pine wood without and with (not shown) fungal decay, choice between fungal wood with and without JHA treatment). When JHA treated filter paper was administered most old soldiers died long before new soldiers were developed. When pine wood was offered, a high percentage of the old soldiers also died, predominantly though, not till the time of the formation of new soldiers. In the other cases the mortality rate was increasingly reduced and was mainly restricted to the first phase of the appearance of new soldiers (see Chap. 4).

It is true that the castes are not determined by the nutrition (Ref 26), however the given examples show that, at least, at times, the development of particular castes can be furthered by adaptation to the actual nutritional situation, and that even extreme demands, such as the application of JHA, can be the

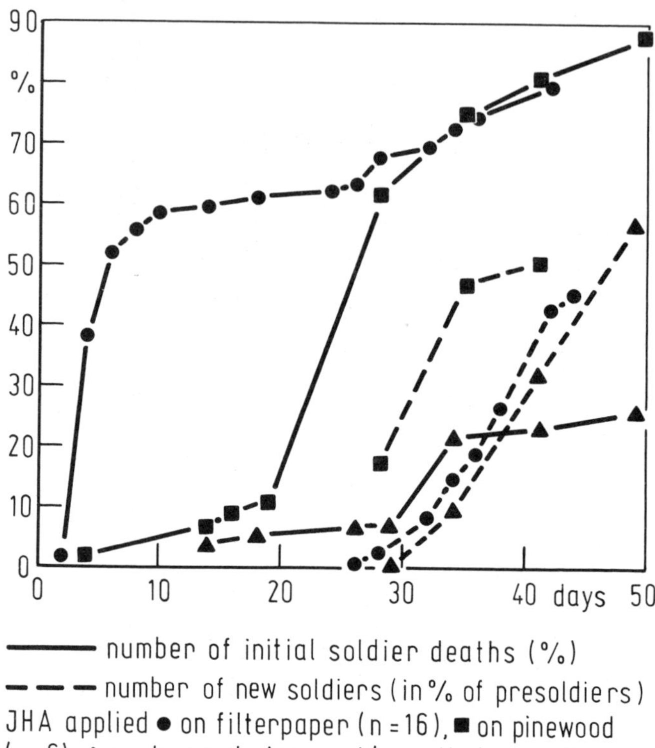

Fig. 1. Mortality rate of existent soldiers in colonies of Kalotermes flavicollis in relation to the formation of new soldiers after application of the JHA Altozar with improved nutritional conditions.

sooner compensated for, the better the health of the termites is. This emphasizes that the physiology of the whole colony determines the fate of the individual (Ref 2,7). Fluctuations in the ratio of castes are correspondingly influenced by the age of the colony and the seasonal variations (Ref 2,6,26,33,34,35, 37).

4. INFLUENCE OF GROUP COMPOSITION ON CASTE FORMATION

The chosen method to apply JHA to definite termite groups, with the subsequent increase in the number of regressive moults, is mainly suitable for the investigation of the regulation of the number of soldiers in a colony. For the similarly complex topics of the formation of neotenics and alates, the numerous investigations on *Kalotermes flavicollis* and the most recently published reviews which also include information on other species, are noted (Ref 17,18,23,26,31,41).

Results of investigations with *K. flavicollis* indicate, that the reproductives in a colony activate the formation of soldiers (Ref 39,40). Similarly, pseudergates stimulate, even if to a lesser extent, other competent members of the group. This influence is intensified with increasing insect numbers (Ref 42, 43). The effect is based upon a rise in the JH titre in individuals by activation of the corpora allata or by the supply of JH, either with the food (trophallaxis, cannibalism) or by intake in vapour form (Ref 26). In the species *Prorhinotermes simplex* (Ref 30) and *K. flavicollis* (Ref 39), existent soldiers retard or inhibit the formation of new soldiers. The old soldiers apparently either produce a JH inhibitor, or degrade JH more easily than the other castes (Ref 23,26). This principle may well have a wider general validity, as was shown in investigations with *Nasutitermes exitiosus* (Ref 10). Taking this into consideration, the proposal of using JHA for the control of termites through the production of excess soldiers, and the subsequent breakdown of colony structures, may prove difficult to realize in practice (Ref 20).

4.1 The Significance of Soldiers on the Differentiation of New Soldiers

4.1.1 Soldier formation in termite species with a low natural proportion of soldiers

Kalotermes flavicollis. Another, until now undescribed, reaction of existent soldiers to the formation of new soldiers is evident in Fig. 1. When presoldiers appeared in large numbers in a colony under the influence of JHA, the mortality rate of old soldiers, in well nourished colonies (see Chap. 3), did not exceed 11%, but it suddenly increased (up to 50% according to the nutritional situation) as soon as the presoldiers moulted to soldiers. In poorly fed groups (filter paper) this procedure began already during the moulting of pseudergates to presoldiers. Evidently the existent soldiers are unable to cope with the continual irritation and demand to produce a JH inhibitor for any great length of time.

K. flavicollis has a relatively small number of soldiers (Ref 13). In the test colonies, 8 - 76 larvae and nymphs occurred per soldier. The capacity of existent soldiers in natural colonies to suppress the differentiation of new soldiers may not be strongly developed, or not often demanded for, as compared to the situation where a juvenoid is fed directly. This was also emphasized by experiments with groups of 15 pseudergates and nymphs in the absence or presence of a soldier (n = 5). The rate of transformation of "workers" to presoldiers was 72 or 58% when Altozar was fed and 61 or 45% when Altosid was given (both JHA's at a concentration of 0.5%; v/v). The respective values for the number of soldiers formed were 50 or 37% and 16 or 14.5%. The presence or absence of one soldier brought no significant differences (χ^2 test). In the controls, isolated presoldiers (3times) and progressive moults (alates, neotenics) were observed.

Working with insects of mature colonies of two species of Zootermopsis, which have a still lower proportion of soldiers as compared to K. flavicollis, it was not possible to detect any inhibiting effect of the soldiers (Ref 24), even though in incipient colonies of Z. angusticollis, the first formed soldier hinders the differentiation of further soldiers (Ref 9). On the other hand, in species with a higher percentage of soldiers, a greater regulating ability should be expected. For example in groups of 45 workers of Nasutitermes exitiosus (natural soldier proportion up to 17.5%; Ref 13), already the presence of one soldier reduces the number of moults to presoldiers (Ref 10).

4.1.2 Soldier formation in termite species with a high natural proportion of soldiers

Coptotermes amanii. There are no available data on soldier proportions in field colonies of this Rhinotermitid. In the laboratory, after 1.5 to 2 years, a ratio of one soldier to 3 workers appeared (unpubl.).

When JHA was fed to groups of 30 or 200 workers, with or without soldiers, the soldiers of C. amanii demonstrated a noticeably better developed ability to hinder the formation of new soldiers, as compared to those of Kalotermes flavicollis (Table 1,2). In both cases, with or without the presence of soldiers, many presoldiers were formed in the larger groups, but significantly fewer differentiated in those groups where old soldiers were present (χ^2; $p<0.001$). In the groups consisting only of workers, the first new soldiers appeared after 18 days. By this time, nearly all the insects were dead in groups which contained workers with additional old soldiers (Table 2). Correspondingly, with a rise in the number of old soldiers in tests with only 30 individuals the survival time was reduced, in contrast to the tendency to suppress the moulting of presoldiers which increased (Table 1). Apparently, through an excess of the substances with JH properties, the inhibiting function of the old soldiers is overactivated, thus implicating the whole group. When the soldiers actually liberate a JH inhibitor, its concentration may rise to noxious proportions for the workers.

In the controls,without JHA treatment (Table 2), the differentiation of pseudergates to presoldiers was also extensively suppressed with the presence of soldiers (χ^2; $p < 0.001$). In comparison to groups composed only of workers, the few presoldier moults occurred with a delay of 19 days.

TABLE 1 Influence of soldiers on the formation of presoldiers in groups of 30 workers of Coptotermes amanii after application of the JHA Altosid

treatment[a]	soldiers added	survival (days)	presoldiers (%)
solvent control	-[b]	> 15.0	-
Altosid[c]	0	13.0	27.3
	3	9.8	1.3
	5	8.4	0

[a] n = 5

[b] no difference for the three group compositions

[c] 0.05 ml of 0.5% solution in acetone (v/v) on discs of filterpaper

TABLE 2 Influence of soldiers on the formation of presoldiers and soldiers in groups of 200 workers of Coptotermes amanii after application of the JHA Altosid

treatment[a]	soldiers added	survival (days)	presoldiers (%)	soldiers (%)
solvent control	0	> 31.0	3.2	1.8
	10	> 31.0	0.2	0
Altosid[b]	0	22.8	48.6	7.4
	10	18.8	38.9	0

[a] solvent control n = 3, Altosid n = 5

[b] s. Table 1, 0.1 ml solution/disc

Nasutitermes nigriceps. After one year of culturing, the soldier proportion was found to be 18% in incipient colonies (unpubl.). For the investigation, the first stage of the larger workers (LW1), which can moult to large soldiers (LS) (Ref 29) was used since not enough individuals were available from the small worker type, from which small soldiers (SS) develop. A greater number of old workers (LW4+5) were also added. LS are fairly frequently found in mounds of N. exitiosus (Ref 28,29). They are also formed in laboratory cultures of N. arborum (Ref 34). The moulting of LW1 to the presoldier stage can also be induced by JHA (Ref 10). Until now, in the large colony of N. nigriceps (Ref 11) only SS have been observed.

In this species it was also possible to obtain presoldiers with the help of JHA (Table 3) and not only from LW1 but also from LW4+5 (in the last case intercastes, worker/presoldier, often form). Apparently, these stages also possess the potential to form soldiers, which is, however, not realized under natural conditions. In the controls, moulting amongst the LW4+5 was observed, but this led only to the formation of workers with lesser head pigmentation. This "regressive" moulting was observed comparatively often in groups which consisted only of LW4+5 (unpubl.). This representative of the Termitidae shows

TABLE 3 Total number of presoldiers for all replicates in groups of workers of Nasutitermes nigriceps after application of the JHA Altozar

groups of 5 LW1 + 25 LW4+5[a)]				groups of 30 LW1 + 200 LW4+5[a)]			
amount of JHA 0.5%[b)] (ml)	soldiers added	presoldiers LW1	presoldiers LW4+5[c)]	amount of JHA 0.5%[b)] (ml)	soldiers added	presoldiers LW1	presoldiers LW4+5[c)]
solvent control	-[d)]	-	-	solvent control	-[d)]	-	-
0.025	0	1	-	0.1	0	1	3
	1	3	2		3	-	-
	3	-	-		10	-	-
0.05	0	-	2	0.2	0	3	1
	1	1	2		3	2	3
	3	-	-		10	-	6

a) LW1, 4+5: large workers of the stages 1,4 and 5 (n = 5)
b) s. Table 1, solution given on sawdust from applewood
c) some individuals were intermediates worker/presoldier
d) no difference for the three group compositions

that here the plasticity of developmental possibilities, the presence of which is characteristic of the Kalo- and Rhinotermitidae, has not yet fully disappeared.

The relatively low presoldier formation rate in N. nigriceps permits only a careful interpretation of the results (Table 3). It seems as if the initially added SS could suppress the differentiation of LW1 to LS even when greater amounts of JHA were given. The moulting of LW4+5 to presoldiers (intercastes), on the other hand, was suppressed to a noticeably lesser extent. In the groups which received treated food, more insects were prepared for moulting to presoldiers than the following counts showed. As a rule, the presoldiers disappeared quickly, too. The moulted insects were apparently actively eliminated since, under the test conditions, the corpses were not, or only partially eaten by the other group members. In N. nigriceps the regulating influence of the soldiers, with respect to the differentiation of new soldiers, is thus intensified by the workers.

Coptotermes niger. Unfortunately no information on the percentage composition of soldiers in natural colonies is available. Estimations from the Berlin laboratory cultures show that the value may be higher than that for C. amanii.

The feeding of JHA to groups of 30 workers, with varying numbers of added soldiers, led to divergent results, as against those of the other species (Table 4). Compared with groups composed of workers only, the addition of a few soldiers (ratio 1:10) stimulated the differentiation of workers to presoldiers. The difference was highly significant (χ^2; $p < 0.001$), also in

TABLE 4 Influence of soldiers on the formation of presoldiers in groups of 30 workers of Coptotermes niger after application of two JHA

amount of JHA 0.5% (ml)[a]	soldiers added	survival (days)	presoldiers (%)
solvent control	-[b]	39.0	-
Altozar 0.025	0	28.6	48.0
	3	32.0	72.0
	5	32.4	59.3
Altosid 0.05	0	14.8	33.3
	3	16.2	76.0
	5	17.2	54.0

a) s. Table 1, n = 5

b) no difference for the three group compositions

repeat tests, at the chosen concentrations of both JHA's. With a higher soldier ratio (1:6), many workers still moulted to presoldiers, though fewer than in the aforementioned case (χ^2; Altozar $p < 0.025$; Altosid $p < 0.001$), but still more than in groups without soldiers (Altozar $p = 0.05$; Altosid $p < 0.001$). Apparently, either the soldiers of C. niger, in contrast to those of the other species investigated, produce the JH inhibitor only at a higher JH concentration, or their ability to degrade JH is not as strongly developed.

Under natural conditions, the situation where no soldier is present in the colony would not occur. A loss of some soldiers could occur, for example after fighting, but this situation would be restabilized as quickly as possible. The speed of the process of differentiation, influenced by the remaining soldiers, is likely to be beneficial to the colony.

4.2 The Significance of Nymphs in the Differentiation of New Soldiers

In various species of the Kalotermitidae and Rhinotermitidae even the late nymph stages are still able to moult to presoldiers and soldiers. Such observations were also made on Zootermopsis nevadensis (Ref 47,48). Experimentally, even the transformation of late nymphs of Kalotermes flavicollis to intercastes soldier/alate can be induced (Ref 19).

Although knowledge of the behaviour of other castes in a termite colony could be drawn from the literature, no data could be derived as to how widely nymphs exert an influence on soldier formation.

Reticulitermes flavipes and R. lucifugus. The addition of 5 large nymphs (equivalent to the nymphs of stages 5 and 6 in R. lucifugus santonensis; Ref 7) and the simultaneous feeding of JHA to groups of 30 workers slightly stimulated the differentiation of pseudergates to presoldiers (χ^2; R. flavipes $p < 0.025$; R. lucifugus $p < 0.05$; Table 5). In the controls the large nymphs frequently moulted to neotenics, but after JHA addition they became presoldiers only. Preliminary experiments showed that the stimulation provided by the nymphs was again counteracted by the addition of soldiers (R. flavipes). Accordingly, in Kalotermes flavicollis the opposite effect of the reproductives as against that of the soldiers was observed (Ref 39).

Heterotermes indicola. When workers were added for their maintenance, large nymphs (probably Ny3+4; so far no investigations on the developmental pathway of the genus Heterotermes have been carried out) showed a surprisingly prompt and intensive reaction to JHA. Depending on JHA type and dosage 64 - 98% of the nymphs transformed into presoldiers and 44 - 94% into soldiers with the simultaneous reduction of their wing buds (both JHA's were tested, each at two concentrations, one example is given in Table 6). Normally within the Heterotermitinae all soldiers are essentially produced only from workers (Ref 12). 10 days after the first actual appearance of nymphal presoldiers already well over 50% of this caste had formed in 3 cases. In the 4th case this per-

TABLE 5 Influence of nymphs on the formation of presoldiers in groups of 30 workers of Reticulitermes flavipes and R. lucifugus after application of the JHA Altozar

termite species	amount of JHA 0.5% (ml)[a]	number of nymphs added	neotenics (%)		presoldiers (%)	
			workers	nymphs	workers	nymphs
Ret. flavipes[b]	solvent control	-	0.7	* *	-	*
		5	-	48.0	-	-
	0.025	-	-	*	28.0	*
		5	-	-	41.3	60.0
Ret. lucifugus[c]	solvent control	-	1.1	*	-	*
		5	-	80.0	-	-
	0.05	-	-	*	16.6	*
		5	-	-	28.6	28.0

a) s. Table 1
b) n = 5
c) solvent control n = 3, JHA n = 5

TABLE 6 Influence of nymphs on the formation of presoldiers and soldiers in groups of Heterotermes indicola after application of the JHA Altozar

amount of JHA 0.1% (ml)[a]	group composition[b]	neotenics (%)		presoldiers (%)		soldiers (%)	
		w	ny	w	ny	w	ny
solvent control	ny	*	-	*	-	*	-
	w	0.6	*	0.8	*	0.4	*
	ny + w	1.4	1.3	0.4	-	0.4	-
0.05	ny	*	-	*	9.3	*	-
	w	-	*	27.6	*	9.2	*
	ny + w	-	-	11.8	98.6	2.8	94.0

a) s. Table 1, n = 5
b) groups composed of either large nymphs (ny), or 100 workers (w), or 30 large nymphs + 100 workers

centage was reached after 14 days; among the workers it was only attained after 16 - 34 days. The quick transformation reveals that the nymphs, independent of their competence phase, require a lower JHA dosage for their differentiation to soldiers than the workers.

The large nymphal soldiers showed the same behavioural pattern as the small soldiers. It is also noteworthy that the moulting was completed without problems and that only a slight loss among the nymphs resulted. When workers, from the hitherto investigated species, transformed into presoldiers and soldiers, insects often died during ecdysis, as was reported for *Reticulitermes lucifugus santonensis* (Ref 16).

In pure nymph groups the differentiation to presoldiers was also induced by JHA, however, most insects died shortly before, or during the moulting. The addition of workers, and therefore proper provisioning, allowed the moulting to proceed without problems.

In the presence of nymphs, the percentage of workers differentiating to presoldiers (χ^2; $p < 0.001$ for the two concentrations of both JHA's) was significantly lower than in nymphless groups (Table 6).

In all replicates JHA was available in excess. The lower transformation rate of the workers in the presence of nymphs may either be related to an inhibiting action of the more responsive nymphs or it could also be conceivable that the workers are less susceptible to JHA when they are occupied with the maintenance of the nymphs. The reactions of the workers would then be adapted to the group's requirements and be thus independent from the intensity of hormonal influences.

The diverging reactions of nymphs of *Reticulitermes* which stimulate soldier formation and nymphs of *Heterotermes* which suppress it, may perhaps be attributable to the differing ages of the nymphs. The used nymphs of *H. indicola* were from an earlier developmental stage. The number of moultings to presoldiers in groups of workers of *R. lucifugus* was compared in the presence of younger or older nymphs. Preliminary investigations revealed that also in this species the young nymphs reduced, or at least did not stimulate the moulting to presoldiers.

In connection with the stimulation of soldier differentiation by late nymphal stages, the observations are of interest in which the maximum proportion of soldiers of several tropical termite species coincides with the swarming period of the alates (Ref 6,37).

5. DISCUSSION

A specific JH titre can be associated with every developmental stage in a termite. In the time between two moults this level may either be raised by the activation of the corpora allata or by supplementing JH from external sources, or it may be reduced by various inhibiting influences or kept constant (Ref 23,26). Unless the developmental cycle of termites is more or less fixed, as is the case in the Termitidae, the possibility for a termite to differentiate into a final caste (alate, neotenic, soldier) is limited to certain periods in the moulting interval. (Ref 25,26,35,42).

As experiments with JHA have shown, these processes are partly influenced by the physiological constitution of the individuals which in turn depends upon seasonal and temperature changes (Ref 26), time of exposure to JH (Ref 15), JH stability, rate of decomposition, and the quality of the available food (Ref 2,20; Chap. 3). Still more important is, however, the sum of the activating and inhibiting influences exerted by the other members of a colony. These influences are also called the "group effects". The aforementioned abiotic and biotic factors also play an important role in the actual behaviour of the group.

In a colony with all specific castes of a certain species, continuous contact among the members and exchange of food guarantees the optimum intake and transfer of pheromones (also JH, see Ref 23,26) which regulate caste determination. The speed with which substances are distributed in a colony by trophallaxis is accelerated by the presence of the royal couple and with the increasing size of the colony (Ref 1,27). The concentration of hormones in an individual and consequently caste formation are also dependent upon this distribution. The differing affinity of the individual castes (Ref 45,46), variation in the function of a termite with its age (Ref 14; see Chap. 4.2) are also important factors. Furthermore, the natural caste proportions, which are species specific, appear to influence the regulating ability of the different developmental stages (see Chap. 4.1 and 4.2). This also means that the actual JH thresholds for the differentiation of the castes have variable levels which are dependent upon the species in question.

6. SUMMARY

The influence of the environmental factor nutrition and termite group composition during the process of caste determination, especially that of soldiers, are examined (feeding of JHA). For example, the number of progressive moults (nymphs, alates, neotenics) increases after feeding on wood decayed by brown rot fungi. In natural colonies of _Kalotermes flavicollis_, the extent to which old soldiers react to the formation of new soldiers is dependent upon the quality of the food. The suppression of new soldiers by soldiers already present in groups, differs in species with a low natural proportion of soldiers (_Zootermopsis_, _Kalotermes_) from those with a higher proportion of soldiers (_Coptotermes amanii_, _C. niger_, _Nasutitermes nigriceps_). The ability of the latter, under the chosen test conditions, to hinder the differentiation of soldiers is more pronounced. The active elimination of surplus soldiers through workers (_N. nigriceps_) or the stimulation of the number of presoldier moults in the presence of a few old soldiers (C. niger) could be observed. In N. nigriceps (Termitidae) even older worker stages still possess the potential to differentiate to soldiers. While older nymphs stimulate the number of presoldier moults (Reticulitermes flavipes, R. lucifugus), younger nymphs tend to inhibit them (Heterotermes indicola). Large nymphs of these species are also able to moult to presoldiers and soldiers. For every developmental stage a certain JH titre is characteristic. Events which may influence the actual JH level in a termite are briefly discussed.

REFERENCES

(1) J. Alibert, Influence de la société et de l'individu sur la trophallaxie chez Calotermes flavicollis Fabr. et Cubitermes fungifaber (Isoptera), L'Effet de Groupe chez les Animaux, Coll. internat. Centre nat. Rech. sci. No.173, 1968.

(2) G. Becker, Über Kastenbildung und Umwelteinfluß bei Termiten, Biol. Zbl. 67, 407-444 (1948).

(3) G. Becker, Versuche über den Einfluß von Braunfäulepilzen auf Wahl und Ausnutzung der Holznahrung durch Termiten, Material u. Organismen 1, 95-156 (1965).

(4) G. Becker, 33 Jahre Termiten-Forschung und -Prüfung in Berlin-Dahlem, Wiss. Ber. aus der Arbeit der BAM, 357-360, 1973.

(5) G. Becker, Termites and fungi, Material u. Organismen, Beih. 3, 465-478 (1976).

(6) P. Bodot, Composition des colonies de termites: ses fluctuation au cours du temps, Insectes sociaux 16, 39-54 (1969).

(7) H. H. R. Buchli, L'origine des castes et les potentialités ontogéniques des termites européens du genre Reticulitermes Holmgren, Ann. Sci. Nat., Zool. 11, 267-429 (1958).

(8) F. L. Carter, Responses of subterranean termites to wood extractives, Material u. Organismen, Beih. 3, 357-364 (1976).

(9) G. B. Castle, The damp-wood termites of western United States, genus Zootermopsis (formerly, Termopsis), Kofoid, C.A., Termites and Termite Control, Univ. Calif. Press, Berkeley, 273-310, 1934.

(10) J. R. J. French, A juvenile hormone analogue inducing caste differentiation in the Australian termite, Nasutitermes exitiosus (Hill) (Isoptera: Termitidae), J. Austral. entomol. Soc. 13, 353-355 (1974).

(11) M. L. Garcia, G. Becker, Influence of temperature on the development of incipient colonies of Nasutitermes nigriceps (Haldemann), Z. angew. Entomol. 79, 291-300 (1975).

(12) P. P. Grassé, C. Noirot, G. Clément, H. Buchli, Sur la signification de la caste des ouvriers chez les termites, Compt. rend. Acad. Sci. 230, 892-895 (1950).

(13) M. I. Haverty, The proportion of soldiers in termite colo-

nies,
Pers. communication
(14) P. E. Howse, On the division of labour in the primitive termite _Zootermopsis nevadensis_ (Hagen), _Insectes sociaux_ 15, 45-50 (1968).
(15) I. Hrdý, Der Einfluß von zwei Juvenilhormonanalogen auf die Differenzierung der Soldaten bei _Reticulitermes lucifugus santonensis_ Feyt. (Isopt.: Rhinotermitidae), _Z. angew. Entomol._ 72, 129-134 (1972).
(16) I. Hrdý, J. Křeček, Development of superfluous soldiers induced by juvenile hormone analogues in the termite, _Reticulitermes lucifugus santonensis_, _Insectes sociaux_ 19, 105-109 (1972).
(17) D. Lebrun, La détermination des castes du termite à cou jaune (_Calotermes flavicollis_ Fabr.), _Bull. biol. Belg. France_ 100, 141-217 (1966).
(18) D. Lebrun, Glandes endocrines et biologie de _Calotermes flavicollis_, _Proc. VI Congr. IUSSI, Bern_, 131-136 (1969).
(19) D. Lebrun, Intercastes expérimentaux de _Calotermes flavicollis_ Fabr., _Insectes sociaux_ 17, 159-176 (1970)
(20) M. Lenz, Die Wirkung von Juvenilhormon-Analoga (JHA) auf _Kalotermes flavicollis_ und _Coptotermes amanii_ (Isoptera: Kalotermitidae, Rhinotermitidae) bei unterschiedlicher Ernährung der Tiergruppen, _Material u. Organismen_, Beih. 3, 377-392 (1976).
(21) M. Lenz, Comparison of soldier induction under the influence of juvenile hormone analogues in 12 termite species (Isoptera: Kalotermitidae, Termopsidae, Rhinotermitidae, Termitidae), In preparation.
(22) M. Lenz, G. Becker, Einfluß von Basidiomyceten auf die Entwicklung von Ersatzgeschlechtstieren bei _Heterotermes indicola_ (Isoptera), _Material u. Organismen_ 10, 223-237 (1975).
(23) M. Lüscher, Environmental control of juvenile hormone (JH) secretion and caste differentiation in termites. _Gen. and comp. Endocrinol. Suppl._ 3, 509-514 (1972).
(24) M. Lüscher, The influence of the composition of experimental groups on caste development in _Zootermopsis_ (Isoptera), _Proc. VII Congr. IUSSI, London_, 253-256 (1973).
(25) M. Lüscher, Die Kompetenz zur Soldatenbildung bei Larven (Pseudergaten) der Termite _Zootermopsis angusticollis_, _Rev. suiss. Zool._ 81, 710-714 (1974a).
(26) M. Lüscher, Kasten und Kastendifferenzierung bei niederen Termiten, Schmidt, G. H., _Sozialpolymorphismus bei Insekten_, Wiss. Verl.ges.mbH, Stuttgart, 694-739, 1974b.
(27) E. A. McMahan, Food transmission with the _Cryptotermes brevis_ colony (Isoptera: Kalotermitidae), _Ann. entomol. Soc. Amer._ 59, 1131-1137 (1966).
(28) E. A. McMahan, Non-aggressive behaviour in the large soldier of _Nasutitermes exitiosus_ (Hill) (Isoptera: Termitidae),

Insectes sociaux 21, 95-106 (1974).
(29) E. A. McMahan, J. A. L. Watson, Non-reproductive castes and their development in Nasutitermes exitiosus (Hill) (Isoptera), Insectes sociaux 22, 183-198 (1975).
(30) E. M. Miller, The problem of castes and caste differentiation in Prorhinotermes simplex (Hagen), Miami Univ. Bull. 15, 3-27 (1942).
(31) E. M. Miller, Caste differentiation in the lower termites, Krishna, K., Weesner, F. M., Biology of Termites, I. Academic Press, New York, London, 283-310, 1969.
(32) R. Nagin, Caste determination in Neotermes jouteli (Banks), Insectes sociaux 19, 39-61 (1972).
(33) C. Noirot, Le cycle saisonnier chez les termites, Proc. 11th internat. Congr. Entomol., Wien, I, 583-585 (1960).
(34) C. Noirot, Formation of castes in the higher termites, Krishna, K., Weesner, F. M., Biology of Termites, I. Academic Press, New York, London, 311-350, 1969.
(35) C. Noirot, Polymorphismus bei höheren Termiten, Schmidt, G.H., Sozialpolymorphismus bei Insekten, Wiss. Verl.ges.mbH, Stuttgart, 740-765, 1974.
(36) W. A. Sands, The association of termites and fungi, Krishna, K., Weesner, F. M., Biology of Termites, I, Academic Press, New York, London, 495-524, 1969.
(37) P. K. Sen-Sarma, S. C. Mishra, Seasonal fluctuations of colony composition and population in Neotermes bosei Snyder (Insecta: Isoptera: Kalotermitidae), J. Ind. Acad. Wood Sci. 3, 43-48 (1972).
(38) R. V. Smythe, J. K. Mauldin, Soldier differentiation, survival, and wood consumption by normally and abnormally faunated workers of the Formosan termite, Coptotermes formosanus, Ann. entomol. Soc. Amer. 65, 1001-1004 (1972).
(39) A. Springhetti, Influenza dei reali sulla differenziazione dei soldati di Kalotermes flavicollis Fabr. (Isoptera), Proc. VI Congr. IUSSI, Bern,267-273 (1969).
(40) A. Springhetti, Influence of the king and queen on the differentiation of soldiers in Kalotermes flavicollis Fabr. (Isoptera), Monitore Zool. Ital. (N. S.) 4, 99-105 (1970).
(41) A. Springhetti, Il controllo dei reali sulla differenziazione degli alati in Kalotermes flavicollis Fabr. (Isoptera), Boll. Zool. 38, 101-110 (1971).
(42) A. Springhetti, The competence of Kalotermes flavicollis Fabr. (Isoptera) pseudergates to differentiate into soldiers, Monitore Zool. Ital. (N. S.) 6, 97-111 (1972).
(43) A. Springhetti, Group effects in the differentiation of the soldiers of Kalotermes flavicollis Fabr. (Isoptera), Insectes sociaux 20, 333-342 (1973a).
(44) A. Springhetti, Il ruolo delle pseudergati nella differenziazione dei soldati di Kalotermes flavicollis Fabr. (Isoptera), Atti Accad. Sci. Ferrara 50, 1-15 (1973b).

(45) A. M. Stuart, The role of chemicals in termite communication,
Johnston, J. W., jr., Moulton, D. G., Turk, A., Advances in Chemoreception, I., Communication by chemical signals,
Meredith Corp., New York, 79-106, 1970.
(46) H. Verron, Rôle des stimuli chimiques dans l'attraction sociale chez *Calotermes flavicollis* (Fabr.),
Insectes sociaux 10, 167-184 (1963).
(47) K. Wanyonyi, The influence of the juvenile hormone analogue ZR 512 (Zoecon) on caste development in *Zootermopsis nevadensis* (Hagen) (Isoptera),
Insectes sociaux 21, 35-44 (1974).
(48) K. Wanyonyi, M. Lüscher, The action of juvenile hormone analogues on caste development in *Zootermopsis* (Isoptera),
Proc. VII Congr. IUSSI, London, 392-395 (1973).

EVIDENCE FOR AN ENDOCRINE CONTROL OF CASTE DETERMINATION IN HIGHER TERMITES

Martin Lüscher
University of Bern, Zoological Institute, Division of Animal Physiology, Engehaldenstrasse 6, CH-3012 Bern, Switzerland

INTRODUCTION

The higher termites of the family Termitidae show a much more rigid caste system than the lower termites of the families Kalotermitidae, Termopsidae and Rhinotermitidae, to mention only those of which caste determination has been studied extensively. While it seems that in lower termites all larvae and nymphs* have a high plasticity and are able to undergo any differentiation including regressive development, determination in higher termites is apparently restricted to a few decisive steps in the development. This is best demonstrated by the diagrams of the development of a lower (*Kalotermes*) and a higher termite (*Macrotermes*) shown in Figs. 1 and 2.

The development of *Macrotermes bellicosus* and of many other higher termites has been worked out by Noirot (3, 4). It can be seen from Fig. 2 that in *Macrotermes* sex is an important factor in caste differentiation, since neuter development is different in males and females. In both sexes the reproductive line of the nymphs separates very early from the neuter (soldier-worker) line. The decision occurs during the first larval instar or in the egg or even during oocyte maturation in the queen. Thus, it may well be that this determination is blastogenic. In the neuter males no further determination is needed, all individuals develop into major workers. The female neuters however undergo two further determination steps in the 3rd larval instars where small presoldiers may develop and in the still unpigmented or only slightly pigmented minor worker, which can still moult into a large presoldier. As in lower termites soldier development is determined when the presoldier development is induced.

We have chosen the example of the fungus grower *Macrotermes*, because we have performed some studies on the possible endocrine mechanisms involved in caste determination in the species *Macrotermes subhyalinus*. The sex relations may, however be

*The term "nymph" is used in termite biology to denote larvae with wing pads, which belong to the line of reproductive development.

quite different in other Macrotermitidae, as shown by Noirot (4): Soldiers are not necessarily only females as in Macrotermes. They may be exclusively males as in Nasutitermes arborum or they may be males and females in equal numbers as in Amitermes evuncifer which in this respect does not differ from lower termites.

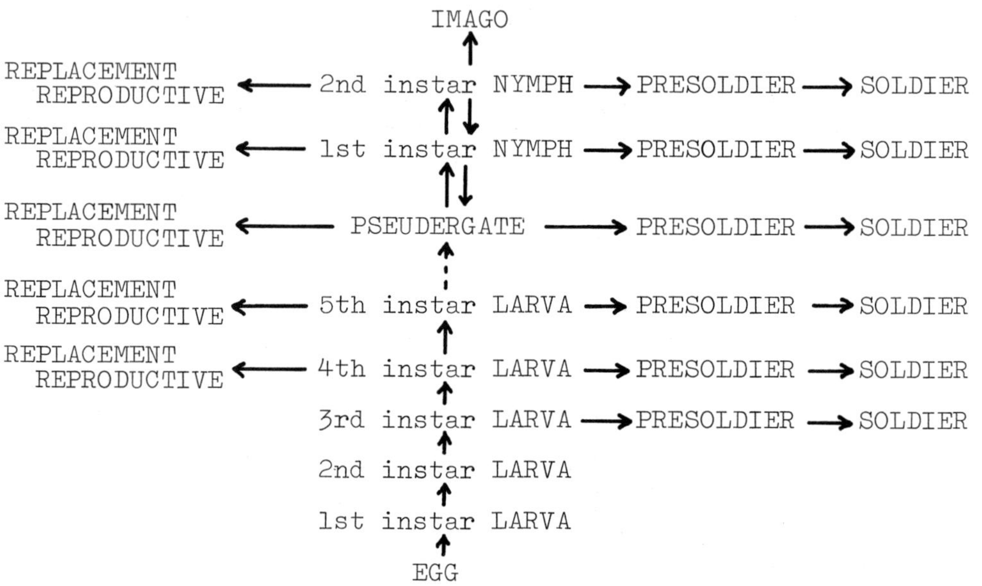

Fig. 1. Developmental possibilities in Kalotermes flavicollis. Each arrow represents a moult. Broken arrows indicate a variable number of moults. After Lüscher (1, 2).

In lower termites caste differentiation is controlled by pheromones, which in the individual termite influence the endocrine system. The hormones then bring about the developmental changes leading to the production of the various castes (review of Lüscher, 1). Assuming similar mechanisms in the higher termites, the existing evidence for pheromonal and hormonal control of caste determination in these more complex societies will now be reviewed.

EVIDENCE FOR A PHEROMONAL CONTROL OF CASTE DIFFERENTIATION

It is well established in lower termites that caste regulation is exerted by inhibitory mechanisms, each caste inhibiting competent larvae and nymphs to differentiate into the same caste (Lüscher, 1). This inhibition is exerted by pheromones which seem to have somewhat antagonistic effects. The pheromones of the reproductive inhibit the development of reproductives, both primary and replacement reproductives and stimulate the development of soldiers. The pheromone of soldiers, however, inhibits soldier development and stimulates reproductive differentiation (Lüscher, 2, 5, 6; Springhetti, 7, 8, 9). But it is not yet certain that the soldiers really give off a pheromone. They may exert their action merely by absorbing or inactivating the pheromones of the reproductives.

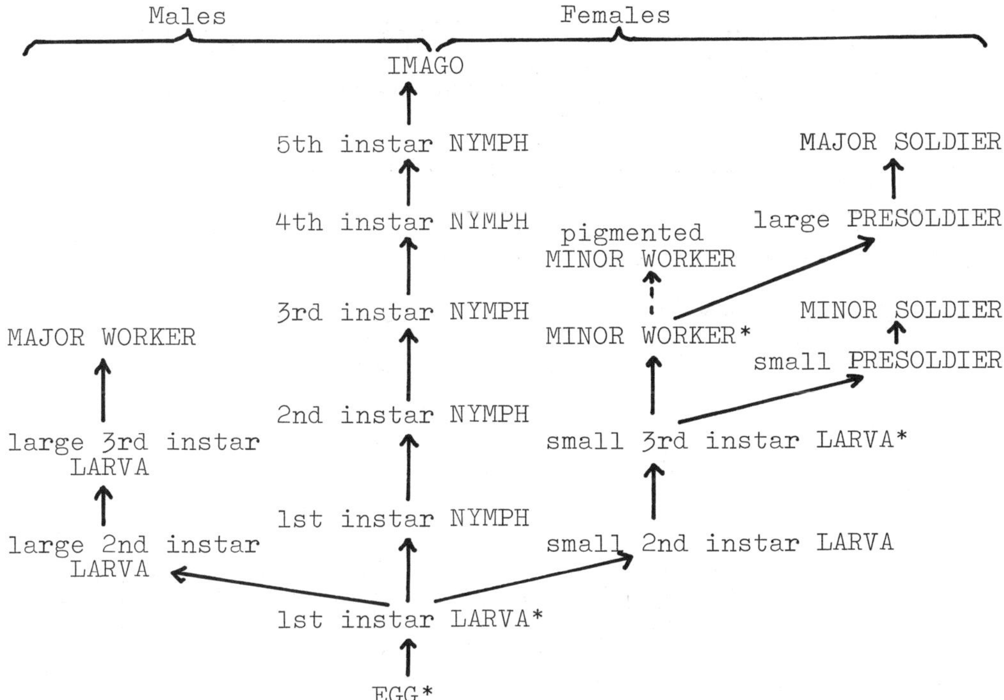

Fig. 2. Developmental possibilities in <u>Macrotermes bellicosus</u>. Each arrow represents a moult. The broken line represents pigmentation without moult. The asterisks indicate the instars in which determination may occur. From Noirot (3), modified.

There is no direct evidence for the existence of pheromones acting on caste determination in higher termites. However, an inhibiting action of the royal pair upon the development of nymphs was demonstrated recently in <u>Macrotermes bellicosus</u> by Bordereau (10). Several colonies were deprived of the royal pair and reopened 2 months later. It was then found that 8 out of 11 colonies contained nymphs of the 2nd, 3rd and 4th instar at a season when normally no nymphs develop.

The same observation was made at least once in *Macrotermes subhyalinus* by J. Darlington (personal communication). The royal pair thus has an inhibitory influence upon nymphal development. This is in agreement with the observation of many authors that nymphal development in most if not in all termites occurs only when the colony has reached a large size, i.e. when inhibitory influences of the royal pair may become diluted enough. This inhibition might be exerted by pheromones. It is possible that soldiers also inhibit soldier development and that replacement reproductives - where they occur - are inhibited in their development by other reproductives. Whether all these inhibitions exist and are exerted by pheromones remains hypothetical.

EVIDENCE FOR A HORMONAL CONTROL OF CASTE DIFFERENTIATION

In 1958 it was discovered that implanted corpora allata from reproductives could induce soldier development in *Kalotermes flavicollis* (Lüscher, 11). Later soldier differentiation could also be induced with synthetic juvenile hormone (Lüscher, 12). We know now that juvenile hormone (JH) and many juvenile hormone analogues (JHA) induce presoldier and soldier development in many species of lower termites and in one representative of the higher termites, *Nasutitermes nigriceps* (see contributions by Hrdy and Lenz in this symposium). It therefore seems very likely that the role of JH in soldier development is a universal phenomenon amongst termites and is not restricted to lower termites. In this context it is interesting that in *Neocapritermes* and *Nasutitermes* the corpora allata increase in volume when soldiers develop (Kaiser, 13). This was also conclusively demonstrated by Wanyonyi (unpublished) for a species of *Odontotermes*, as shown in Table 1.

TABLE 1 Corpora Allata Volume of Odontotermes spec.

	n	
Larvae	6	16'700 ± 2'700 μ^3
Workers	6	10'750 ± 1'490 μ^3
Presoldiers	5	261'090 ± 33'240 μ^3
Soldiers	1	20'510 μ^3

Odontotermes is closely related to *Macrotermes* in which it can now be assumed that an activation of the corpora allata in female 3rd instar larvae leads to minor soldiers and that an activation in young female workers leads to the development of major soldiers. It remains unknown, however, what the external or internal influences are that cause or prevent this activation.

It is reasonable to assume that the decision about the further development of the 1st instar larva is also dependent on JH, since in the lower termites the development of nymphs can be influenced by JHAs.

The corpora allata volume of 1st instar larvae has not yet been investigated, because such an investigation seems at first sight not very promising as long as we cannot determine the age within the instar accurately. Furthermore, an extensive investigation over the seasons of the year would be necessary in order to possibly find at one particular season a certain proportion with less voluminous or less active glands. Even at the season when development of 1st instar nymphs occurs, there is still a much larger proportion of the 1st instar larvae moulting into 2nd instar larvae which later become workers and soldiers.

If we assume that determination at this stage is blastogenic and occurs already in the egg during its maturation in the ovary of the queen, mainly two possibilities seem to exist. As in certain ants the eggs may differ in the nutritional value of their yolk or different amounts of JH may be incorporated in the yolk. There is no evidence for differences in the nutritional value of the yolk or in its quantity since up to now no significant differences in the volume of freshly laid eggs could be detected. However, remarkable differences in the JH content of eggs from different mounds of *Macrotermes subhyalinus* could be demonstrated (Lanzrein, Abo-Khatwa and Lüscher, unpublished).

Eggs from various mounds near Kajiado, Kenya were collected and cleaned from soil particles. About 1 gram (corresponding to approximately 10'000 eggs) from each mound was extracted in ether-ethanol, purified by TLC according to the method described by Lanzrein et al. (14) and assayed for JH-content using the *Galleria* wax test. The results are expressed in *Galleria* units (GU) per fresh weight. They are given in the diagram in Fig. 3.

JH was detected in all of the 22 samples. The samples show an extremely marked variation. Most of the samples fall into two groups of less than 2 and more than 7 GU/mg. The occurrence of these two activity levels may have a meaning for the later development, and if so, it is conceivable that only those eggs with a low JH titre may be competent for the development of nymphs. Since we know that nymphs of *M. subhyalinus* near Kajiado appear in the mounds in April and May, it may be expected that the JH content of the eggs should be particularly low during the months January to March. We had no samples collected in March, but the values for samples collected in January and February are considerably lower than those of samples collected during the other seasons, as shown in Table 2.

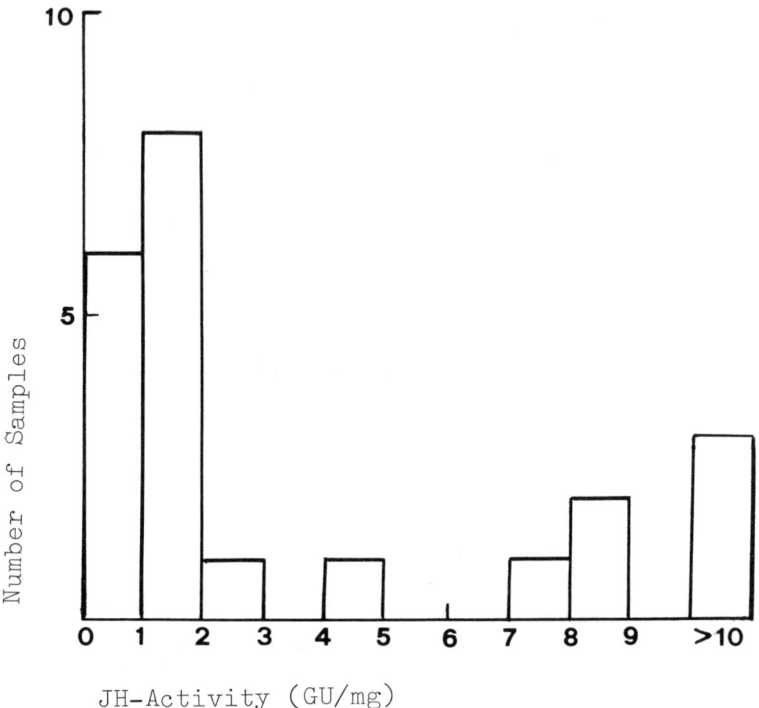

Fig. 3. Diagram of the distribution of JH contents of eggs from 22 mounds of Macrotermes subhyalinus.

TABLE 2 JH Titre of Eggs of Macrotermes subhyalinus Collected During Different Seasons

Period	n	JH titre GU/mg	% with less than 3 GU/mg
Jan./Feb.	11	3.5 ± 1.8	82 %
April to December	11	9.3 ± 5.0	55 %
Whole year	22	6.6 ± 2.7	68 %

There are not yet enough values from each season to allow a thorough statistical analysis, but the values obtained so far seem to indicate in general a lower JH content of the eggs one or two months before the appearence of nymphs. This suggests a partly blastogenic determination or predetermination by incorporation of varying amounts of JH into the eggs during yolk formation in the ovary of the queen.

If the eggs show variable contents of JH and if these variations are correlated to the seasons, they should also be evident in the haemolymph of the queen. The JH titre of queen haemolymph samples of various seasons was therefore determined with the same methods.

Some JH could be expected in the queen of Macrotermes since the corpora allata are very much enlarged and their volume can be about 100 times that of a female imago (Table 3 and Fig. 4).

TABLE 3 Corpora Allata Volume of Adults of Macrotermes subhyalinus (fresh tissue volume)

Female imago	7.5 mio	μ^3
King	39.5×10^6	μ^3
Queen (small, 4.7 g)	252×10^6	μ^3
Queen (large, 15.7 g)	665×10^6	μ^3

JH activity could be expected in connection with the enormous production of eggs (up to 40'000 per day), although the mechanisms of the production of vitellogenin and of its incorporation into the oocytes seems to be different from that in another related insect like for instance the cockroach Nauphoeta. While in the latter vitellogenin is synthesized under the influence of JH in the fat body (Bühlmann, 15) and transferred also under the influence of JH to the oocytes (Wilhelm and Lüscher, 16), most of the proteins in the Macrotermes queen are synthesized in the ovary rather than in the fat body (Wyss-Huber and Lüscher, 17). Another remarkable difference between the cockroach Nauphoeta and the Macrotermes queen was found in the oxidative metabolism of fatbody and ovary homogenates (Abo-Khatwa and Lüscher, unpublished). In the fat body homogenate of the adult female cockroach the oxidation rate of succinate is accelerated significantly by the addition of JH III and also by β-ecdyson while cockroach ovary homogenates do not respond to these hormones. Fat body homogenates of termite queens, however, do not react to either of the hormones while queen ovary homogenates respond to β-ecdyson with an instant and significant increase in the respiration rate (up to 50%). JH had in this case an antagonistic effect to β-ecdyson stimulation. All these findings seem to indicate a completely different hormonal regulation of oocyte maturation in the termite queen, ecdyson being the hormone stimulating oocyte maturation and protein synthesis in the ovary, while JH may not be necessary for this or may even have an inhibitory effect.

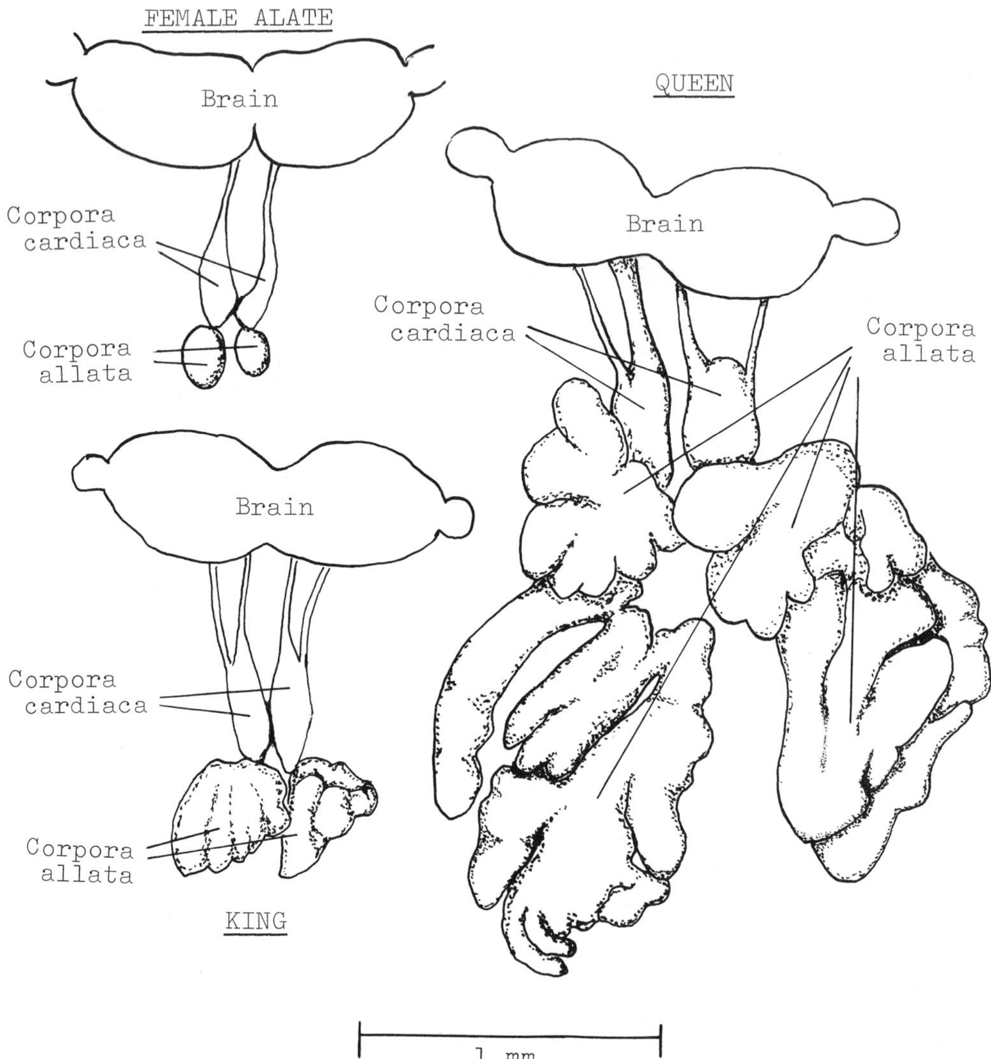

Fig. 4. Dissected and fixed endocrine systems of adult M. subhyalinus, drawn from photographs.

In fact ecdysone was recently detected in large amounts in the queen of Macrotermes bellicosus by Bordereau et al. (18). These authors found as much as 2150 ng per g fresh weight in the ovary and 65 ng per g in the haemolymph. Similar results were obtained independently in our laboratory by Lanzrein, O'Connor and Lüscher (unpublished) for the queen of M. subhyalinus. These findings and those mentioned above suggest that ecdyson like in Aedes (Hagedorn et al., 19) is the main oocyte maturation hormone. It is therefore somewhat surprising that the haemolymph of the queen of M. subhyalinus contains extremely high titres of JH. The distribution of the JH titres of 27 samples collected during various seasons is shown in Fig. 5.

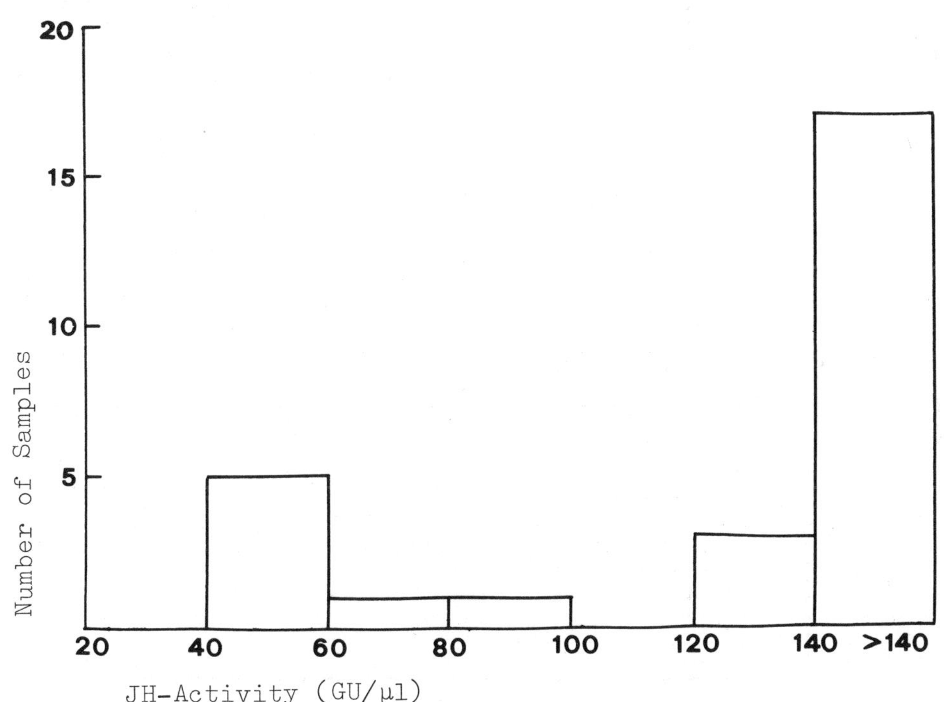

Fig. 5. Diagram of the distribution of JH contents of haemolymph of 27 queens of M. subhyalinus.

Although also in this case we do not yet have enough values to allow statistical analysis, the results are suggestive. There seems to be a period of low activity (Aug.-Nov.) which preceeds the period of low activity of the eggs. Since this period of low activity is about 3 months earlier than the one in the eggs, it seems likely that JH is incorporated very early during oogenesis into the oocytes. It must be borne in mind that the production of nymphs and alates, although bound to a particular season, is very variable. The percentage of 1st instar larvae becoming nymphs varies but is always rather low. Exact figures cannot yet be given, but an investigation of the rate of nymphal development in dependence of other factors is in progress (J. Darlington). It has already revealed that in some, even large mounds there may in certain years be no nymphs developing at all. Therefore a 100% low JH titre cannot be expected at any season, even if our hypothesis is correct.

This being assumed, it is interesting to note that the JH titre of female _Macrotermes_ dealates rises at a very rapid rate after colony foundation, at the time when the first eggs are laid (Abo-Khatwa, unpublished). The results of this investigation are shown in Fig. 6. This sudden and rapid rise may besides promoting oocyte maturation which at this age may still be regulated in the "normal way" by JH, lead to a high incorporation of JH in the eggs, which in young colonies never develop into nymphs.

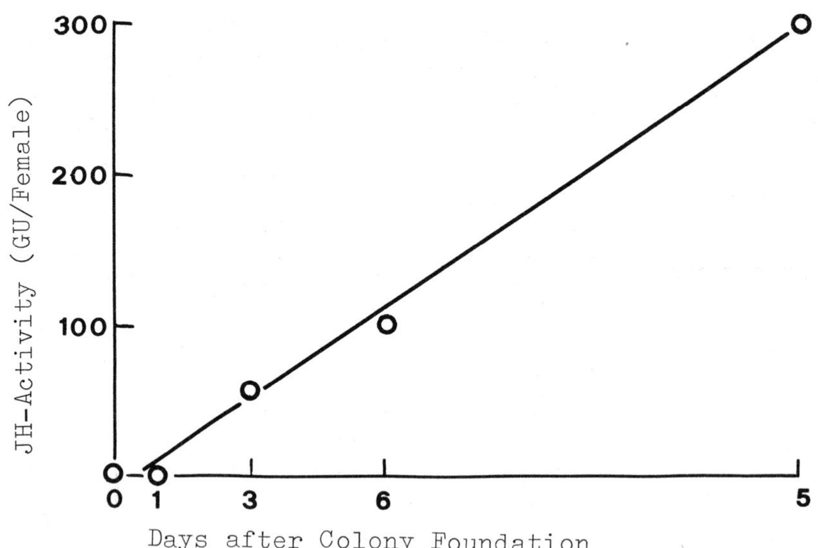

Fig. 6. JH titre of dealate females of _M. subhyalinus_ after colony foundation (Abo-Khatwa, unpublished).

If the neuters are determined by a high JH titre of the egg, we must, however, assume that this blastogenic determination is not irreversible and can still be influenced during the first instar of larval life, as shown by the experiments of Bordereau (10) leading to the development of nymphs after removal of the royal pair. This seems to indicate that also during the first larval instar JH or a corpora allata stimulating pheromone is given off by the royal pair and taken up directly or indirectly from workers by the already predetermined 1st instar worker larva.

If juvenile hormone is being fed to larvae by the workers, there are mainly 2 sources from where they can get it. One source can be the anal fluid which is regularly given off by the queen in very large amounts. It is attractive to the workers and is eagerly taken up by them. It does contain JH (Meyer and Lüscher, 21) but only occasionally in large quantities. The other source may be the eggs. It is well known that termites like many other social insects show oophagy and workers eating eggs can sometimes be observed. With the eggs containing amounts of JH which vary with the seasons, the workers could take up similarly varying amounts of hormone and transmit them to the larvae. Unfortunately we have no precise data on the frequency of oophagy and also on the frequency of workers feeding the larvae, so that it is impossible to evaluate the rate of hormone transmission achieved in this way. Oophagy is generally only observed in small colonies and it is doubtful if it occurs at all in large ones.

Another possibility of larvae getting JH is through their normal food which is supposed to be the substratum of the fungus comb after it has been changed by the metabolism of the fungus *Termitomyces*. Sannasi et al. (22) have reported quite high JH activity in extracts of fungus combs and Conidia of several species of *Odontotermes*, detected with the *Tenebrio*-test. We were unable to confirm these findings for *M. subhyalinus*, in which no JH activity could be detected in fungus combs or Conidia.

CONCLUSIONS

The evidence presented here seems to indicate that similarly to lower termites, a representative of higher termites, *Macrotermes*, makes use of hormones, especially JH for regulating caste development. Caste determination is probably regulated in the first place by external factors like seasonal influences which may be temperature, humidity, physiological condition of food plants etc., but also by pheromones given off by members of the termite colony and circulating by means of trophallaxis. These external factors may act on the endocrine system and at this point the evidence becomes more convincing. JH titres change with the seasons in queen haemolymph and in eggs laid some time later. The JH content of the egg seems to be a predetermining factor deciding about neuter larva or reproductive nymph determination in the 1st instar larva. However, some influence from the royal pair is still evident at this stage. This may be due to pheromones or JH given off by the queen. Later in the development of the neuters the decision about the development of soldiers or workers in the female larvae is probably again brought about by JH titre differences. It follows from corpora allata volume measurements that these glands become very active when the larva is to become a presoldier and later a soldier. The corpora allata are probably under the influence of an inhibiting soldier pheromone and of stimulating reproductive pheromones.

No proof of the assumptions presented here can be given, because experimentation is almost impossible as long as the higher termites cannot be kept under controlled conditions in the laboratory. We have now succeeded in keeping young colonies of Macrotermes subhyalinus in the laboratory in apparently good condition for 19 months. The colonies contain now 500-1000 individuals. They have developed all castes - major and minor workers as well as major and minor soldiers - but no nymphs have yet been observed. We hope that with these colonies it will be possible to perform experiments which will yield more conclusive evidence relating to the mechanisms involved in caste determination.

REFERENCES

(1) M. Lüscher, Kasten und Kastendifferenzierung bei niederen Termiten, in G. H. Schmidt, Sozialpolymorphismus bei Insekten, Wissenschaftl. Verlagsges. Stuttgart, 694-739 (1974).

(2) M. Lüscher, Pheromones and polymorphism in bees and termites, in IUSSI-Symposium Pheromones and Defensive Secretions, Univ. Dijon, 123-141 (1975).

(3) Ch. Noirot, Formation of castes in the higher termites, in K. Krishna and F. Weesner, Biology of Termites, Academic Press, 1, 89-123 (1969).

(4) Ch. Noirot, Polymorphismus bei höheren Termiten, in G. H. Schmidt, Sozialpolymorphismus bei Insekten, Wissenschaftl. Verlagsges. Stuttgart, 740-765 (1974).

(5) M. Lüscher, Environmental control of juvenile hormone (JH) secretion on caste differentiation in termites, Gen., Comp. Endocrinol., Suppl. 3, 509-514 (1972).

(6) M. Lüscher, The influence of the composition of experimental groups on caste development in Zootermopsis (Isoptera), Proc. VII Congr. IUSSI, London, 253-256 (1973).

(7) A. Springhetti, Influenza dei reali sulla differenziazione dei soldati di Kalotermes flavicollis Fabr. (Isoptera), Proc. VI Congr. IUSSI, Bern, 267-273 (1969).

(8) A. Springhetti, Influence of king and queen on the differentiation of soldiers in Kalotermes flavicollis Fabr. (Isoptera), Monitore Zool. Ital. (N.S.), 4, 99-105 (1970).

(9) A. Springhetti, Il controllo dei reali sulla differenziazione degli alati in Kalotermes flavicollis Fabr. (Isoptera), Bolletino di Zool., 38, 101-110 (1971).

(10) C. Bordereau, Déterminisme des castes chez les termites supérieurs: mise en évidence d'un contrôle royal dans la formation de la caste sexuée chez Macrotermes bellicosus Smeathman (Isoptera, Termitidae), Insectes Sociaux, 22, 363-374 (1975).

(11) M. Lüscher, Experimentelle Erzeugung von Soldaten bei der Termite *Kalotermes flavicollis* (Fabr.), *Naturwiss.* 45, 69-70 (1958).
(12) M. Lüscher, Die Bedeutung des Juvenilhormons für die Differenzierung der Soldaten bei der Termite *Kalotermes flacollis*, *Proc. VI Congr. IUSSI, Bern*, 165-170 (1969).
(13) P. Kaiser, Die Hormonalorgane der Termiten im Zusammenhang mit der Entstehung ihrer Kasten, *Mitt. Hamburg. Zool. Mus. u. Inst.* 54, 129-178 (1956).
(14) B. Lanzrein, M. Hashimoto, V. Parmakovich, K. Nakanishi, R. Wilhelm and M. Lüscher, Identification and quantification of juvenile hormones from different developmental stages of the cockroach *Nauphoeta cinerea*, *Life Sciences* 16, 1271-1284 (1975).
(15) G. Bühlmann, Vitellogenin in adulten Weibchen der Schabe *Nauphoeta cinerea* - Immunologische Untersuchungen über Herkunft und Einbau, *Rev. suisse Zool.* 81, 642-647 (1974).
(16) R. Wilhelm and M. Lüscher, On the relative importance of juvenile hormone and vitellogenin for oocyte growth in the cockroach *Nauphoeta cinerea*, *J. Insect Physiol.* 20, 1887-1894 (1974).
(17) M. Wyss-Huber and M. Lüscher, Protein synthesis in "fat body" and ovary of the physogastric queen of *Macrotermes subhyalinus*, *J. Insect Physiol.* 21, 1697-1704 (1975).
(18) C. Bordereau, M. Hirn, J.-P. Delbecque et M. de Reggi, Présence d'ecdysones chez un insecte adulte: la reine de termite. *C.R. Acad. Sci. Paris* 282, 885-888 (1976).
(19) H. H. Hagedorn, J. D. O'Connor, M. S. Fuchs, B. Sage, D. Schlaeger and M. K. Bohm, The ovary as a source of α-ecdysone in an adult mosquito, *Proc. Nat. Acad. Sci. USA* 72, 3255-3259 (1975).
(20) D. R. Meyer, B. Lanzrein, M. Lüscher and K. Nakanishi, Isolation and identification of a juvenile hormone (JH) in termites. *Exper.* (in press).
(21) D. Meyer and M. Lüscher, Juvenile hormone activity in the haemolymph and the anal secretion of the queen of *Macrotermes subhyalinus* (Rambur) (Isoptera, Termitidae), *Proc. VII Congr. IUSSI, London*, 268-273 (1973).
(22) A. Sannasi, P. K. Sen-Sarma, C. George and S. Basalingappa, Juvenile hormone activity from various sources of termite castes and their fungus gardens, *Insectes Sociaux* 19, 81-86 (1972).

JUVENILE HORMONE AND APHID POLYMORPHISM

Dinah F. Hales*
School of Biological Sciences, Macquarie University, North Ryde, N.S.W. 2113 Australia.

ABSTRACT

Evidence for the role of juvenile hormone in determination of apterous parthenogenetic aphids is reviewed. Corpus allatum activity, assessed by volume or nuclear size, is greater in adult aphids producing apterous young, and in larval aphids which are developing as apterae. On the basis of these relationships, a consistent hypothesis for the control of production of apterae by juvenile hormone can be proposed. However, experiments with juvenile hormone analogues have given equivocal or negative results and the hypothesis remains unconfirmed. Some of the difficulties involved in experimental work of this kind are indicated.

INTRODUCTION

Aphids show several kinds of polymorphism, related to an annual cycle which allows maximum exploitation of temporarily favorable host plants. It may be helpful to give a simplified summary of the annual cycle and the terminology used to refer to some of the morphs. Full reviews of aphid polymorphism can be found in Lees (1966) and Hille Ris Lambers (1966).

In most aphids showing a complete annual cycle (holocyclic species) fertilized eggs are laid in autumn and hatch in spring giving rise to females called fundatrices, which are nearly always apterous, viviparous and parthenogenetic. They in turn give birth to further parthenogenetic females, and parthenogenetic viviparous reproduction usually continues until autumn, both alate and apterous females occurring. In the decreasing temperatures and photoperiods of autumn, sexual males and sexual oviparous females are produced. These mate and the overwintering eggs are laid. The cycle may involve migration from a woody primary host, on which the eggs are laid, to herbaceous secondary hosts. In species or races with this kind of cycle, the first response to shortening photoperiod is the production of a physiologically distinct winged morph, the sexupara, which returns to the primary host and gives birth to wingless oviparae. Alate males born on the secondary host then migrate to mate with the oviparae. Some strains of aphids apparently lose the ability to produce sexual forms after they have been reared for many generations in long day conditions. Such strains are termed anholocyclic. In the absence of sexual reproduction, the

*Née White.

main type of polymorphism to be considered is the dichotomy between the alate and apterous parthenogenetic females, and it is the physiological control of these alternative pathways of development which will occupy the rest of this paper.

Apterous parthenogenetic aphids differ from alate parthenogenetic aphids in numerous ways, both in morphology and physiology. Apart from the obvious lack of wings and associated thoracic and muscular development, apterous aphids have fewer antennal sensoria, no ocelli, sometimes reduced compound eyes, shorter appendages and usually reduced body pigmentation and sclerotisation. On the other hand, they have a shorter larval period and a greater reproductive potential. It has recently been observed that between the alate and apterous extremes, there may occur intermediates, e.g. aphids with the external appearance of alates but the physiological and behavioural characteristics of apterae (Shaw 1970, Burns 1972). However, there have been no studies on the endocrine control of production of these intermediates.

Generally speaking, the apterous parthenogenetic aphid can be considered as a neotenic, larva-like form, while the alate is a more typical adult insect and probably preceded the apterous form in the phylogeny of the group. The determination of an individual aphid as alate or apterous is brought about by environmental factors acting either just before or just after birth. The majority of work on aphid polymorphism, especially before about 1960, was concerned with defining the environmental triggers which induced the production of different morphs; the most important factors controlling the alate/apterous dimorphism were generally considered to be temperature, nutrition and crowding, although there were inconsistencies between the results of different workers. Work along these lines continues but will not be reviewed here. The desirability of a physiological (rather than an "ecological") approach to the mechanism of form determination was recognised long ago. Shull (1937) proposed several mechanisms by which a hypothetical hormone could influence form. Evans (1938) suggested that the corpus allatum might be involved. Lamb (1956) proposed a system whereby ecdysone in conjunction with juvenile hormone might control form, noting that the differences between alate and apterous adults were like differences between larvae and alate adults. The abnormal development of parasitised aphids was attributed to upsets in the balance between juvenile hormone and ecdysone by Johnson (1959). Lees (1961, 1966) suggested that the corpus allatum might be involved in the production of apterae, and showed that juvenile hormone analogues could suppress the development of alate characteristics when they were applied to third and fourth instar alatiform larvae. Von Dehn (1963) had already shown that farnesol could cause postnatal determination of apterae in _Aphis fabae_ Scop. although Hille Ris Lambers (1966) criticised these experiments on the grounds that the crowding factor might not have been adequately considered. Of the authors mentioned above, only Lees and von Dehn were able to supply experimental evidence to support their proposals; hormone analogues were not available for testing by earlier workers. Lees (1975) has recently reviewed some aspects of the physiological control of aphid polymorphism, laying stress on neuroendocrine components, and has now concluded that juvenile hormone is not involved in alate/apterous dimorphism. The evidence is discussed later in this paper.

Even in recent years, there have been rather few studies on the physiological control of alate/apterous dimorphism in aphids. The present author's studies on the cabbage aphid, Brevicoryne brassicae (L.) can be used as a basis for discussion. The approach taken initially in these studies was to re-investigate some of the environmental factors influencing production of alatae or apterae in the cabbage aphid and to study changes in the apparent activity of the corpus allatum in different environmental conditions (estimated from the volume, or later the nuclear diameter). The apparent activity of the corpus allatum could then be correlated with environmental conditions producing alatae or apterae.

ONTOGENY OF ALATE AND APTEROUS FORMS

Before looking at environmental factors producing apterae or alatae, the sequence of developmental events involved will be briefly summarized.

Determination of Alate or Apterous Development

Parthenogenetically reproducing aphids are viviparous, and an individual commences ovulating while still an embryo itself. Environmental factors influencing form are effective either on the late embryo (via the mother or directly) or in the first few days of larval life, although there is some variation between species. The period of sensitivity to external triggers is sharply defined in some species, but extended in others. In the cabbage aphid, I found that most late embryos and young larvae have histologically recognisable wingbuds, which regress when strong apterising stimuli are applied, either before or after birth (White 1971). On this basis it can be supposed that all or most embryos start out with the potential for becoming alate, and may be diverted from this course by apterising stimuli. This is in accordance with the developmental relationships in aphids proposed by Johnson and Birks (1960). Other authors consider that the "normal" course of development is either neutral or towards the apterous form, and that alatae develop only in response to positive alata-promoting stimuli (Lees 1966, Johnson 1966, Applebaum et al. 1975).

The implication of an initially "neutral" developmental pathway is, of course, that two kinds of physiological switches exist: one determining alate development and one determining apterous development. The two switches could in theory be the presence or absence of adequate concentrations of a particular substance (? hormone) or they could be more distantly related. However it should be remembered that an individual can be determined irreversibly as an aptera early in the sensitive period, and alata-promoting stimuli are thereafter ineffective. On the other hand an individual which has retained the potential for developing as an alate may still be caused to develop as an aptera by stimuli applied as late as the second larval stage in some species. The response is thus asymmetrical.

The developing wingbuds are first recognisable in living aphids in the third larval stage. Morphometric differences between alatae and apterae can also be observed at this time.

Changes in the Corpus Allatum during Development

If differentiation of alatae and apterae is mediated by juvenile hormone, one would expect to find differences in the apparent activity of the corpus allatum (c.a.) in the two forms during development. Such differences were found in the cabbage aphid, using two separate criteria: c.a. volume, and diameter of nuclei in the c.a. (Figs 1, 2) (White 1965, 1968a). By either criterion apteriform larvae had more active corpora allata in the last larval stage, and also in the second-last larval stage, though this was assessed by volume only. It is not, of course, possible to distinguish potential alatae and apterae in earlier larval stages, around the time of determination of the developmental pathway to be followed. However it seems that differences in c.a. function do appear around the time of determination (see below).

Elliott (1975) using Aphis craccivora Koch confirmed the relationship between c.a. activity and form. He showed furthermore that the higher c.a. activity in apteriform larvae correlated well with the precocious ovarian growth in this morph.

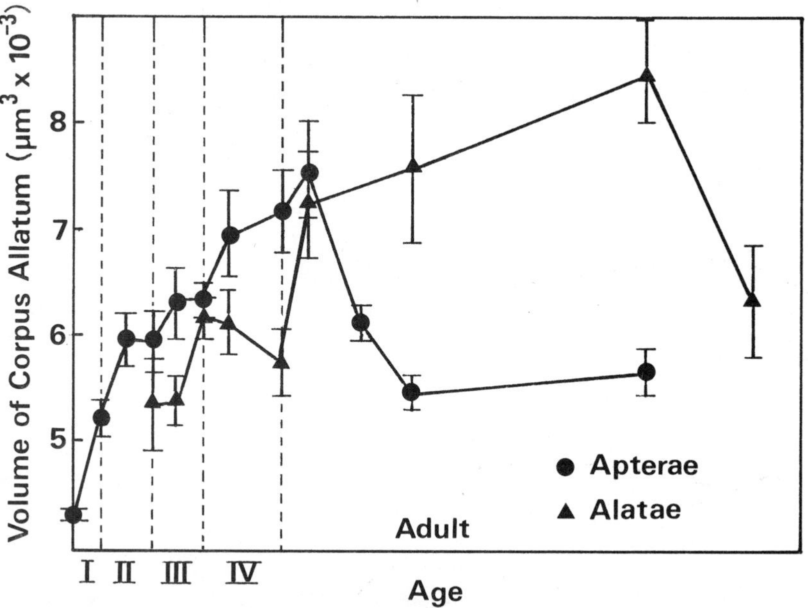

Fig. 1. Changes in volume of corpus allatum with age in the cabbage aphid. The symbols I, II, III, IV refer to the four larval stages.

ENVIRONMENTAL FACTORS INFLUENCING PRODUCTION OF APTEROUS CABBAGE APHIDS

Bonnemaison (1951), in a major work, elucidated many of the main factors involved in form determination in aphids, using Brevicoryne brassicae, Myzus persicae and Sappaphis plantaginea. Any subsequent work on aphid polymorphism, especially in these species, owes a great deal to his contributions.

Form of parent

It is well established that alate aphids usually produce only apterous young except sometimes at the end of their reproductive life. This is true of the cabbage aphid, so one might expect to find differences between the corpora allata of alate and apterous adults. Figure 1 shows that this is indeed the case. The corpora allata of alates appear to be more active than those of apterae when assessed by volume, except just after the adult edcysis. A similar result is obtained when c.a. activity is assessed by nuclear size, but the nuclei in alates are already enlarged at the adult ecdysis (Fig. 2). (White 1965, 1968a; Lamb and White 1971).

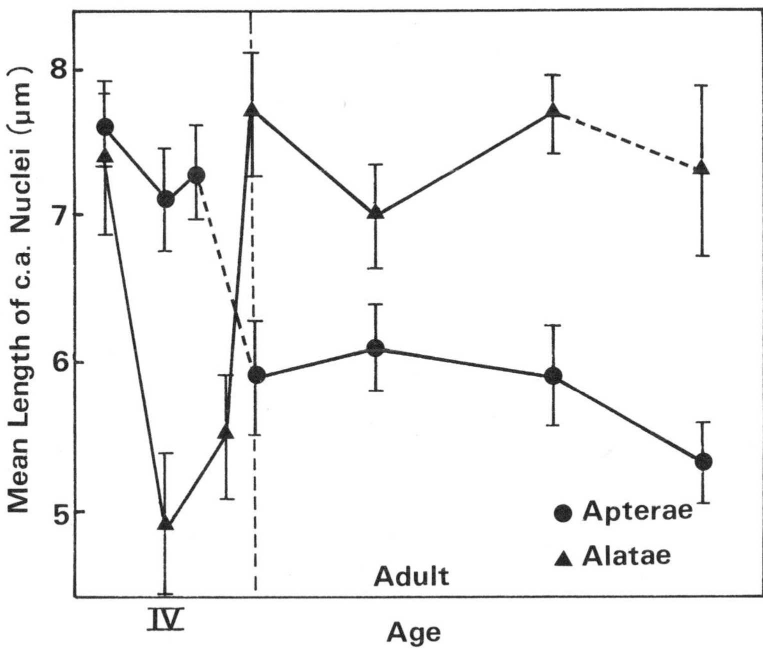

Fig. 2. Changes in length of corpus allatum nuclei in last larval stage and adults of the cabbage aphid.

This apparently high activity explains why many embryos in alate mothers lose their wingbuds before birth (White 1971). Reduced volume and signs of degeneration in the c.a. of old alates correlates also with the incomplete loss of wingbuds in their embryos and their occasional production of alate progeny. The changing activity of the c.a. through reproductive life may account for differential ability of first and last-born progeny of alate parents to respond to crowding and other factors influencing form, by producing alate young themselves. Sutherland (1970) has invoked an "interval timer" to explain such differences. Further work is required to decide whether differences in parental c.a. activity can form the basis of the "interval timer" effect.

Isolation

Bonnemaison (1951) showed that contact between aphids or "effet de groupe" was the major environmental factor stimulating production of alatae in the cabbage aphid. Conversely the present author's investigations showed that when isolated from birth, cabbage aphids invariably developed as apterae (White 1968b). Isolated apterous parents also produced apterous progeny. Isolated apterous adults had more active corpora allata than did crowded apterae. A difference in c.a. activity was already evident by the middle of the first larval stage when newborn larvae were isolated or grouped; the isolated individuals had higher c.a. activity. Similar results were obtained by Kawada (unpublished) with *Aphis craccivora*, alate parents and isolated apterous parents (i.e. parents producing apterous young) having larger corpora allata than grouped apterous parents.

Temperature

Although there is some conflict in the literature (reviewed by Bonnemaison 1951) more recent workers have generally concluded that high temperatures favour the production of apterae (e.g. Lees 1967, Johnson 1966, Lamb and White 1966). However compared with the ability of alate parentage, or of isolation, to produce 100% apterae, temperature gives variable results and is a much less powerful or predictable influence on form determination. It has been shown that apterous aphids kept at 10°C have less active corpora allata than aphids kept at 25°C (White 1968a); this again correlates with the relative production of apterae by the two groups.

Because of the overwhelming influence of isolation on determination of apterae in most aphids, it is difficult to design experiments which are capable of unequivocally distinguishing the effects of less powerful physiological or environmental factors. One is forced to use groups of aphids, and there can be no certainty that the apparent response to the factor under consideration is not really an indirect response, brought about by increased or decreased interaction between the individuals in the group. Even taking this into account, the effect of temperature seems likely to be authentic, as one would expect more activity and interaction at high temperatures, leading to production of more alatae, and this is the reverse of the observed result.

Nutrition

The remaining important factor influencing alate/apterous dimorphism in aphids is food, and the response to food appears to have two components (Mittler and Sutherland 1969, Harrewijn 1972). These components can be referred to as a gustatory response on the one hand, and a nutritional response on the other.

The gustatory response is seen when aphids are placed on seedling host plants which provide food of high quality; they immediately begin to produce apterous young although they may previously have been producing alatae. Since the response is so rapid it is unlikely to be caused by alterations in nutritional status, but is probably due to sensory stimuli which activate appropriate neuroendocrine pathways, finally resulting in determination of apterae. This pattern of stimulus and response appears to be analogous to the response to isolation, where the absence of tactile stimuli results in production of apterae. No experiments correlating the gustatory response with c.a. or other endocrine activity have been done.

The nutritional response is less well-defined. Aphids placed on nutritionally inferior artificial diets produce more apterae over a period of time than those feeding on better artificial diets (as indicated by growth rate, adult size etc.) Although deficient diets cause production of apterae they also cause reduced c.a. activity (White 1972); similar correlations of starvation and c.a. activity have been shown in a number of insects. Production of apterae in these conditions may not be mediated by endocrine mechanisms; it may simply reflect an inadequate supply of metabolites for differentiation of alate structures. The same kind of effect is seen when specific amino acids (isoleucine or histidine) are omitted from the diet (Mittler and Dadd 1966) and when specific vitamins (folic acid) are made unavailable (Raccah et al. 1973). In all these cases the dietary deficiency may prevent differentiation of "luxury structures" such as wings directly, as insufficient precursors are available for DNA and protein synthesis. Dietary deficiencies probably have little to do with the general mechanism of control of alate/apterous dimorphism.

CONTROL OF PRODUCTION OF APTERAE BY JUVENILE HORMONE - A WORKING HYPOTHESIS

The close correlation between c.a. activity and determination of apterae during ontogeny and in most experimental conditions allows us to build up a hypothesis of the control of alate/apterous dimorphism by c.a. function (White 1968a, 1971). According to the hypothesis, environmental stimuli such as isolation cause increased c.a. activity in adults, and the increased juvenile hormone concentration causes regression or non-development of alate characteristics in their embryos. The initial high juvenile hormone concentration has a positive feedback effect, enhancing the c.a. activity of these embryos throughout their subsequent larval life, and continuously repressing development of alate structures. Environmental stimuli can also affect young aphids for a short period after birth, with similar results. Riddiford (1971) has since provided incontrovertible evidence for the positive feedback effect of juvenile hormone applied to the embryo on the activity of the corpus allatum throughout larval development in *Pyrrhocoris*.

The indirect evidence for the hypothesis is thus quite consistent. One would think it might be easy to confirm it by use of the wide range of juvenile hormone analogues now available. Unfortunately this does not seem to be the case; the few such experiments that have reached the literature have had negative or dubious results. They are briefly discussed below.

EXPERIMENTS WITH JUVENILE HORMONE ANALOGUES

The experiments of von Dehn (1963) with farnesol have already been mentioned.

White and Lamb (1968) showed an increased proportion of apterae in the offspring of adult apterous cabbage aphids treated with 2 μg of a juvenile hormone analogue (JHA). This dose was sufficient to increase mortality among both adults and young, so the possibility of a reduced crowding effect cannot be completely ruled out, although every effort was made to maintain equal numbers in each treatment cage.

Treatment of newborn cabbage aphids (White 1968c) showed a small difference between doses of JHA but the result can be considered negative because of the relatively high proportion of apterae in the control group.

Steel and Lees have been unable to influence determination of apterae in Megoura viciae by applications of juvenile hormone analogues (Steel, personal communication).

Applebaum et al. (1975) failed to find effects of juvenile hormone analogues on determination of apterae in Myzus persicae. Unfortunately they provided little information on the conditions of their experiments and stock cultures; the remarkably uniform numbers of alatae in their experiments could perhaps have been photoperiodically induced sexuparae.

Application of JHA to third or fourth instar aphid larvae results in juvenilisation, as it does in other insects. Such experiments throw no light on the mechanism of form determination which occurs much earlier in life.

Experiments primarily designed to test the effectiveness of juvenile hormone analogues as "third generation pesticides" are unlikely to give any useful information about a possible role of juvenile hormone in determination of apterae, as many important variables are uncontrolled. The work of Tamaki (1973) however implies some effect, since spraying with a juvenile hormone analogue was said to delay the appearance of alate Myzus persicae.

Lees (1975) mentions unpublished experiments in which he treated presumptive gynoparae (sexuparae producing only female offspring) with JHA and was unable to reverse their normal winged condition. He was, however, able to produce apterization in these individuals by subjecting them to long photoperiods. The system was chosen to ensure that all individuals tested were potential alates, which is not possible in the mixed progenies that occur in parthenogenetic reproduction.

With the same objective, I recently tested a JHA on Myzocallis annulatus, which normally produces only alatae viviparae. No effect on form determination was observed at the doses used.

DISCUSSION

Why have these experiments not given more positive results? Firstly, the interpretation of the corpus allatum data may be incorrect: an enlarged gland or enlarged nuclei may not indicate increased secretion of juvenile hormone. Histological criteria for assessing insect endocrine activity have been discussed by Novak (1966). It is desirable to confirm histological estimates of glandular activy by other means; for example, by implantation of supposedly active glands. Unfortunately the classical surgical techniques of endocrine research are scarcely feasible in aphids. Even injection causes a relatively enormous wounding effect. In the circumstances we are forced to rely on histological assessments of gland activity and it is encouraging to find that the observations are repeatable by different workers, and that they accord well with more easily confirmed results in larger insects (e.g. the low c.a. activity in partially starved aphids mentioned previously). If the histological evidence has been correctly interpreted, c.a. activation is one step in the process of form determination. We must then enquire why determination of apterae cannot be achieved more unequivocally by hormone analogues. In some cases problems of experimental design may be involved. It is necessary to control all variables as far as possible, since the response is labile and can be affected by a number of individual factors and a range of interactions between environmental conditions. In particular, variables must be controlled at a level that ensures that alatae will be produced in the controls. This necessitates working with groups of aphids; immediately the possibility of a differential "effet de groupe" between treatments is introduced.

So far, there has been no attempt to isolate or identify the natural juvenile hormone of any aphid. There is some evidence that earlier analogues were not particularly active on aphids; they required rather high doses on a body-weight basis to produce morphological effects (White and Lamb 1968). In addition, one must note the observations of Brooks (1973) and of Slade and Wilkinson (1973) who have shown that JH analogues act by preventing enzymatic breakdown of natural juvenile hormone. In the absence of juvenile hormone, then, JH analogues cannot be expected to have much effect. The pre-condition for experiments on determination of apterae is to shut off the supply of natural juvenile hormone, and promote the development of alatae. Thus it may not be surprising that JH analogues fail to influence form determination; the natural juvenile hormone concentration is already too low, even though further breakdown may be inhibited. Elliott and McDonald (1976) have shown in addition that activity of the c.a. is further inhibited by postnatal application of JHA. For these reasons I cannot agree with Staal (1972) and Lees (1975) who have concluded that JH is not involved in determination of apterae. Certainly the case is not proven; either the corpus allatum evidence or the juvenile hormone analogue evidence is misleading, and one can present arguments against the rationale of either line of investigation.

If c.a. activation is involved in form determination it is presumably only one step in a series of sensory, neural, neuroendocrine and endocrine events. Lees' experiments (cited above) on the effects of long daylength on potential gynoparae imply control at an earlier step in the series. He has also pointed out other instances of photoperiodic induction of alatae other than sexuparae and has suggested that there may be common elements in the physiological control of production of alate sexuparae and ordinary alatae (see Lees 1975 for references). It may thus be more useful to look for evidence of control at a neuroendocrine level, though this may still in turn affect c.a. secretion.

Any advances in the understanding of alate/apterous dimorphism will depend on development of new approaches and techniques. An anti-juvenile hormone or c.a. blocking agent would be a useful tool for experiments. On another line, Harrewijn (1973) has developed an apparatus for minimising crowding effects in aphid experiments on artificial diets. He is using the system to study the role of indole alkylamines in form determination, suggesting that these substances may constitute a link in the chain between juvenile hormone and morphogenesis.

REFERENCES

Applebaum, S.W., B. Raccah and R. Leiserowitz, Effect of juvenile hormone and β ecdysone on wing determination in the aphid Myzus persicae, J. Insect Physiol. 21, 1279-1282 (1975).

Bonnemaison, L.E., Contribution à l'étude des facteurs provoquant l'apparition des formes ailées et sexuées chez les Aphidinae, Annls Epiphyt. 2, 1-380 (1951).

Brooks, G.T., Insect epoxide hydrase inhibition by juvenile hormone analogues and metabolic inhibitors, Nature (Fri.) 245, 382-384 (1973).

Burns, M., Effect of flight on the production of alatae by the vetch aphid Megoura viciae, Entomologia exp. appl. 15, 319-323 (1972).

von Dehn, M., Hemmung der Flügelbildung durch Farnesol bei der schwarzen Bohnenlaus, Doralis fabae Scop., Naturwiss. 17, 578-9 (1963).

Elliott, H.J., Corpus allatum and ovarian growth in a polymorphic paedogenetic insect, Nature, Lond. 257, 390-391 (1975).

Elliott, H.J. and F.J.D. McDonald, Effect of a juvenile hormone analogue on morphology, reproduction and endocrine activity of the cowpea aphid, Aphis craccivora Koch. J. Aust. ent. Soc. 15, 1-5 (1976).

Evans, A.C., Physiological relationships between insects and their host plants. I. The effect of the chemical composition of the plant on reproduction and production of winged forms in Brevicoryne brassicae (L.) (Aphididae), Ann. appl. Biol. 25, 558-572 (1938).

Harrewijn, P., Wing production by the aphid Myzus persicae related to nutritional factors in potato plants and artificial diets, Proc. Int. Conf. Insect and Mite Nutrition, 575-588 (1972).

Harrewijn, P., Functional significance of indole alkylamines linked to nutritional factors in wing development of the aphid Myzus persicae, Entomologia exp. appl. 16, 499-513 (1973).

Hille Ris Lambers, D., Polymorphism in Aphididae, A. Rev. Ent. 11, 47-78, (1966).

Johnson, B., Effect of parasitisation by Aphidius platensis Brèthes on the developmental physiology of its host Aphis craccivora Koch, Entomologia exp. appl. 2, 82-99, (1959).

Johnson, B., Wing polymorphism in aphids IV. The effect of temperature and photoperiod, Entomologia exp. appl. 9, 301-313 (1966).

Johnson, B., and P.R. Birks, Studies on wing polymorphism in aphids I. The developmental process involved in the production of the different forms, Entomologia exp. appl. 3, 327-339 (1960).

Lamb, K.P., Physiological relations between aphids and their host plants, Ph.D. Thesis, University of Cambridge (1956).

Lamb, K.P. and D. White, Effect of temperature, starvation and crowding on production of alate young by the cabbage aphid (Brevicoryne brassicae), Entomologia exp. appl. 9, 179-184 (1966).

Lamb, K.P. and D.F. White, Endocrine aspects of alary polymorphism in Brevicoryne brassicae (L.), Endocrinologia experimentalis 5, 19-22 (1971).

Lees, A.D., Clonal polymorphism in aphids, Symp. R. ent. Soc. Lond. 1, 68-79 (1961).

Lees, A. D., The control of polymorphism in aphids, Adv. Insect Physiol. 3, 207-277 (1966).

Lees, A.D., The production of the apterous and alate forms in the aphid Megoura viciae Buckton, with special reference to the role of crowding, J. Insect Physiol. 13, 289-318 (1967).

Lees, A.D. Aphid polymorphism and "Darwin's Demon", Proc. R. ent. Soc. Lond. 39, 59-64, (1975).

Mittler, T.E. and R.H. Dadd, Food and wing determination in Myzus persicae (Homoptera:Aphidae) Ann. ent. Soc. Am. 59, 1162-1166 (1966).

Mittler, T.E., and O.R.W. Sutherland, Dietary influences on aphid polymorphism, Entomologia exp. appl. 12, 703-713 (1969).

Novak, V.J.A., Insect Hormones, Methuen, London (1966).

Raccah, B., S.W. Applebaum and A.S. Tahori, The role of folic acid in the appearance of alate forms in Myzus persicae, J. Insect Physiol. 19, 1849-1855 (1973).

Riddiford, L.M., Juvenile hormone and insect embryogenesis, Mitt. Schweiz. Ent. Ges. 44, 177-186 (1971).

Shaw, M.J.P., Effects of population density on alienicolae of Aphis fabae Scop. III, the effect of isolation on the development of form and behaviour of alatae in a laboratory clone, Ann. Appl. Biol. 65, 205-212 (1970).

Shull, A.F., The production of intermediate winged aphids with special reference to the problem of embryonic determination, Biol. Bull. Woods Hole, 72, 259-287 (1937).

Slade, M., and C.F. Wilkinson, Juvenile hormone analogs: A possible case of mistaken identity? Science 181, 672-674 (1973).

Staal, G.B., Biological activity and bio assay of juvenile hormone analogs, in Insect Juvenile Hormones, eds. J.J. Menn and M. Beroza, Academic Press, New York (1972).

Sutherland, O.R.W., An intrinsic factor influencing alate production by two strains of the pea aphid Acrythosiphon pisum, J. Insect Physiol. 16, 1349-1354 (1970).

Tamaki, G., Insect developmental inhibitors: Effect of reduction and delay caused by juvenile hormone mimics on the production of winged migrants of Myzus persicae (Hemiptera, Aphididae) on peach trees, Can. Ent. 105, 761-765 (1973).

White, D., Changes in size of the corpus allatum of a polymorphic insect, Nature, Lond. 208, 807 (1965).

White, D.F., A Physiological Study of Wing Polymorphism in the Cabbage Aphid, Brevicoryne brassicae (L.), Ph.D. thesis, University of Sydney (1968a).

White, D.F., Cabbage aphid: effect of isolation on form and on endocrine activity, Science 159, 218-219 (1968b).

White, D.F., Post-natal treatment of the cabbage aphid with a synthetic juvenile hormone, J. Insect Physiol. 14, 901-912 (1968c).

White, D.F., Corpus allatum activity associated with development of wingbuds in cabbage aphid embryos and larvae, J. Insect Physiol. 17, 761-773 (1971).

White, D., Effect of varying dietary amino acid and sucrose concentrations on production of apterous cabbage aphids, J. Insect Physiol. 18, 1241-1248 (1972).

White, D.F. and K.P. Lamb, Effect of a synthetic juvenile hormone on adult cabbage aphids and their progeny, J. Insect Physiol. 14, 395-402 (1968).

NEUROSECRETORY CONTROL OF POLYMORPHISM IN APHIDS

Colin G.H. Steel
Department of Biology, York University,
Downsview, Ontario M3J 1P3, Canada

ABSTRACT

The production of sexually or asexually reproducing progeny (oviparae or virginoparae) by *Megoura viciae* is maternally determined by daylength. Parent aphids subjected to a change in daylength will switch from production of one morph to the other. This ability is destroyed following selective ablation of one of the five groups (Group I) of neurosecretory cells (NSC) in the brain. Ablation of other groups of NSC in the brain or massive lesions in much of the protocerebrum do not have this effect. The secretory product of these cells is apparently not released into the haemolymph but is transported intra-axonally to the abdomen, where it may be delivered directly to the reproductive system which receives fine branches from the abdominal nerve. Group I NSC are necessary for the production of virginoparae under long day conditions for when destroyed, all female progeny are oviparae. Hence, the active factor from the cells is virginopara-promoting.

The NSC possess collaterals in the cerebral neuropile. Deep lesions which damage these collaterals but not other segments of the NSC prevent switches of morph production and result in depletion of secretory product from the cells. It is inferred that the NSC depend on input from other neurons to regulate their secretory behaviour. This input does not derive from the visual apparatus since destruction of the eyes or optic lobes is without affect. Destruction of a small group of neurons in the lateral anterior protocerebrum prevents switches of morph production in the same way as does damage to the collaterals. It is inferred that these lateral neurons contain a neuronal photoperiodic clock which regulates release of secretion from the NSC via synapses with their collaterals in the neuropile.

It is suggested that a possible action of the maternal secretion may be to manipulate programming of the endocrine system of the progeny and that its local delivery to the reproductive system may serve to prevent interactions with the parent's own endocrine system.

INTRODUCTION

Previous studies of the control of polymorphism in aphids have been concerned primarily with identification and analysis of the environmental factors which influence the appearance of the various morphs in aphid populations (see Lees, 1966, 1972, 1973). Although it was recognized at an early stage that morph determination must involve the endocrine system (Lees, 1963), there have been few studies of the endocrine physiology of aphids. The presence in some adult morphs of certain morphological features characteristic of nymphal stages gave rise to the view that polymorphism was under the control of juvenile hormone. The current evidence relating to this view is discussed elsewhere in this volume by Hales.

However, in species in which polymorphism is controlled maternally, such as the bean and vetch aphid *Megoura viciae* (Lees, 1959), the control of developmental switching appears to be more complex.

The developing female embryos in the ovarioles of viviparous parthenogenetic females (virginoparae) pass through two successive developmental stages in each of which they are switched into one of two developmental pathways, according to sensory cues perceived by the parent. At the first stage, the embryo is determined as either a virginopara or an egg-laying ovipara. At the second stage, embryos determined as virginoparae at the first stage become determined as apterous or alate. Daylength perceived by the mother controls switching at the first stage (Lees, 1959) and the frequency of contacts with other aphids ('crowding') is the principal controlling factor at the second (Lees, 1967). Since daylength and 'crowding' cues may occur independently of each other but may operate simultaneously on embryos of different ages, it appears impossible that one maternal hormone could control developmental switching at both stages.

Indeed, recent evidence obtained with *Megoura* indicates that juvenile hormone may not be involved at either stage of this maternal switching mechanism. It appears that the determination of virginoparae as alate or apterous according to the degree of 'crowding' may be mediated by the fused thoracic ganglionic mass (Lees, personal communication). The determination of embryos as virginoparae or oviparae (at the first switching stage) in response to daylength is the subject of the present article. Evidence is presented that a virginopara-promoting neurosecretory product is elaborated in the brain and transported intra-axonally to the abdomen where it may be delivered directly to the reproductive system. Its release under long day conditions results in the determination of embryos as virginoparae; under short day conditions the secretion is withheld and oviparae result.

THE EXPERIMENTAL SYSTEM

Apterous virginoparae reared at 15°C in 12L, 12D ('short' days, or SD) give birth to oviparae but no virginoparae. If transferred to 16L, 8D ('long' days, or LD) after a prior exposure to SD, the younger embryos within the parent become differentiated as virginoparae, the older embryos being already determined as oviparae by the preceding SD. The parent thus gives birth to an initial sequence of oviparae followed by a sequence of virginoparae. Each parent produces about 100 female progeny. Since the emrbyos become determined sequentially as they reach the critical growth stages, the sequence of progeny produced reflects the past photoperiodic treatment of the parent. Thus, parent virginoparae respond to a change in daylength from SD to LD (or vice-versa) by switching from the production of oviparae to virginoparae (or vice-versa). This ability to switch morph production in response to a change in daylength has been used to assay the integrity of the morph determination system following surgical interference with components of the neuroendocrine system. In most experiments, parent virinoparae were reared to the imaginal ecdysis in SD, thereby determining their older embryos (born first) as oviparae. Operations were performed on the newly ecdysed adult which was then placed in LD for the remainder of adult life and its progeny collected serially to establish if it was able to respond to LD by producing virginoparae. In normal insects under these conditions, the switch-over to virginopara production occurs about half-way through reproductive life. Further details of the photoperiodic switching regimes and the criteria for assessment of integrity of the morph determination mechanism in operated aphids are given elsewhere (Steel and Lees, 1976).

The operations consisted of placing small localised lesions in selected regions of the neuroendocrine system using a brief pulse from a specially designed radio-frequency microcautery. The pulse was delivered through specially fabricated electrolytically tapered tungsten electrodes (Steel, 1976b) inserted directly through the head capsule. Preliminary experiments revealed that both visual inspection and whole mount preparations of the lesioned area gave misleading impressions of the location and extent of tissue damage. Serial sections cut at 5 μm were therefore prepared for all operated insects once their progeny sequence had clearly showed whether or not the operation had affected the ability of the maternal morph determination mechanism to respond to the change in daylength. The sections were stained with paraldehyde fuchsin (PAF) as described elsewhere (Steel, 1976a) in order to obtain information concerning the cytology and distribution of surviving neurosecretory cells (NSC). Over 250 operated aphids have been examined in this way.

Detailed criteria by which electrocoagulated or otherwise damaged areas of tissue are differentiated from intact tissue are given elsewehere (Steel, 1976b).

CEREBRAL NEUROSECRETORY CELLS AND MORPH DETERMINATION

In 1963 Lees suggested that neurosecretion might play a role in the photoperiodic control of polymorphism in *Megoura*. Indirect evidence came from experiments designed to localise the photoreceptors of photoperiodic information (Lees, 1964) in which small regions of parent aphids kept in SD were subjected to supplementary illumination sufficient in duration to extend the daylength to LD over small areas of the body. It was found that the only region of the parent sensitive to such supplementary illumination was the head; further, the most sensitive area was the centre of the dorsum. It was inferred that the brain possessed direct photosensitivity. The remote location of the photoreceptive area from the responding embryos in the abdomen prompted the suggestion that communication between the brain photoreceptors and the embryos might be achieved by a neurosecretory mechanism.

An experimental approach to this problem first required knowledge of the number and distribution of neurosecretory cells in the brain of *Megoura* (Steel, 1976a). This information is summarised in Fig. 1. Five anatomically distinct groups of PAF positive neurons are recognised in the protocerebrum. The relationships

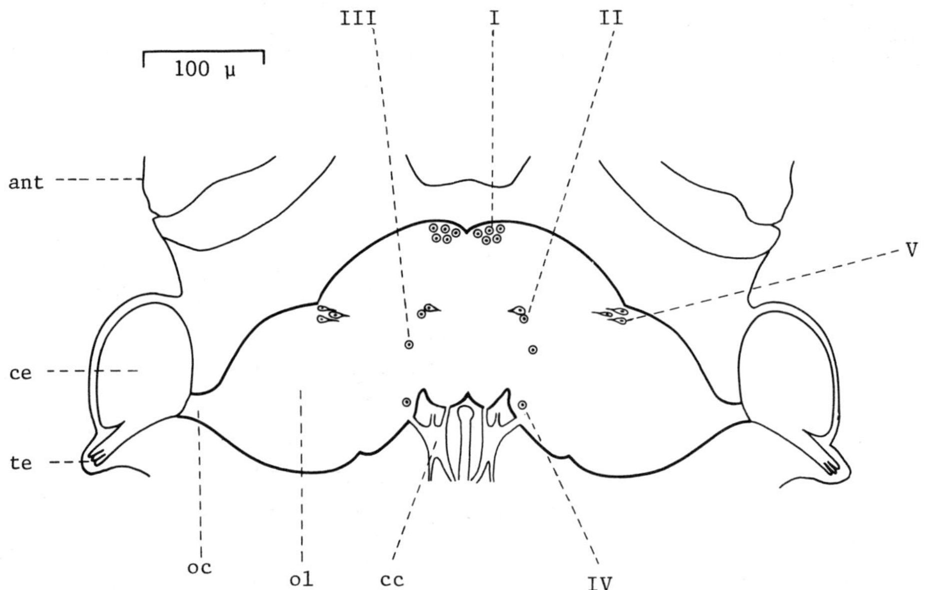

Fig. 1. Dorsal plan view of the head of adult apterous virginopara of *Megoura*, showing distribution of neurosecretory cells. I, II, III, IV, V, Groups of neurosecretory cells; ant., antenna; cc, corpus cardiacum; ce, compound eye; oc, optic chiasma; ol, optic lobe; te, triommatidial eye.

of the axons of these various cells suggested that Groups I and II together comprised the equivalent of the 'medial' NSC group of other insects (Steel, 1976a); however, selective lesioning experiments showed that the functions of larval moulting and tanning of the cuticle were controlled by Group II (Steel, 1976b) and no function for Group I was found.

The possible involvement of one or more of these five groups of NSC in the photoperiodic control of polymorphism was then examined by their selective ablation, either individually or in various combinations. The technique by which the perikarya of neurons can be destroyed without detectable damage to underlying neuropile has been described elsewhere (Steel and Lees, 1976).

Destruction of Group II in adult aphids does not impair a normal response of the morph determination mechanism to a change in day-length. Indeed, massive superficial lesions to the neurons of the dorsal protocerebrum, some of which destroyed NSC Groups II, III and IV bilaterally had no effect on the response (Fig. 2, area C). Longevity and fecundity in these animals were normal. The only abnormality observed was a reduced tendency towards spontaneous locomotor activity. Thus, it is clear that surgical damage to the brain does not in itself interfere with the photo-periodic morph determination mechanism.

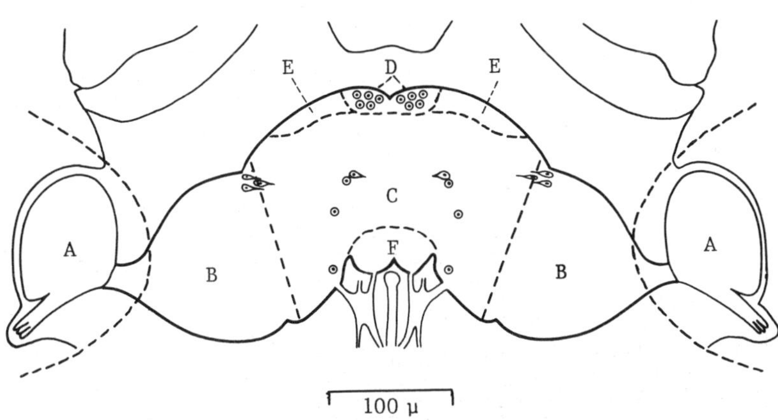

Fig. 2. Dorsal plan of the head of *Megoura* to show the main types of lesion experiments employed. Explanation of lettered areas in the text.

Very small lesions which destroyed all the NSC of Group I (Fig. 2, area D) invariably resulted in a failure of the mechanism, only oviparae being produced in LD. It proved technically impossible to destroy all 10 NSC of Group I without causing some damage to other neurons among and between the NSC, thereby leaving open the possibility that the essential components of this small area of protocerebrum were these other neurons rather than the NSC. This possibility was resolved by comparison of the effects of a number of lesions which were localised mainly in the medial anterior portion of area C (Fig. 2)(which has been shown above to be dispensable) but which extended anteriorly into area D to varying extents. Regardless of the number of immediately adjacent neurons which were destroyed, the mechanism continued to respond normally when as many as eight of the 10 Group I NSC were destroyed. But when all 10 NSC were destroyed the mechanism failed regardless of the extent of damage to adjacent neurons. It is therefore concluded that the essential component of area D in the protocerebrum consists of the Group I NSC.

These experiments provide the first direct evidence that the morph determination mechanism contains a neurosecretory component. Subsequent work was devoted to elucidation of the functional significance and mode of action of this neurosecretory component.

IS THERE A MATERNAL HORMONE?

The interpretation of Group I NSC as equivalent to part of the 'medial' NSC group of other insects (Steel, 1976a) suggested the possibility that Group I might be the source of a neurohormone liberated from the corpora cardiaca which influenced (directly or indirectly) the differentiation of embryos in the abdomen.

Deep lesions were placed in the posterior region of the protocerebrum (Fig. 2, area F) which coagulated the tracts of axons leading to the corpora cardiaca. These lesions resulted in depletion of extrinsic stainable neurosecretory material from the corpora cardiaca, without direct damage to the corpora cardiaca, thus confirming that the tracts had been destroyed. All these animals responded normally to a change in daylength. This experiment also supports the view, stated above, that the corpus allatum of the mother is not involved in the determination of progeny, since the neural regulation of corpus allatum activity is achieved primarily by way of these same nerves.

To investigate the possibility that the putative hormone was released from some neurohaemal site other than the corpora cardiaca, attempts were made to detect it in the haemolymph by means of parabiosis experiments. Joining aphids in parabiosis proved a delicate operation, but in the few successful operated pairs, no change in morph production was encountered (Lees, personal communication). Thus, no evidence of a blood-borne factor influencing polymorphism has been obtained.

An explanation of these negative results became apparent when the detailed axonal projections of Group I NSC were traced. Many aphid neurosecretory cells had been known for some time to be characterised by unusually large accumulations of PAF positive neurosecretory material along their axons (Johnson, 1963). The axons of Group I NSC in *Megoura* were also found to display this feature (Steel, 1976a), making their course through the nervous system relatively easy to trace by light microscopy.

The cell bodies are unipolar, each giving rise to an axon which passes ventrally and posteriorly into the cerebral neuropile, where substantial accumulations of stainable secretory material are found (Fig. 3). In the region of these accumulations, each axon gives off a collateral which passes along the dorsal surface of the central body and then terminates in the neuropile. The axon of each NSC passes beneath the central body and descends into the circum-oesophageal connectives, without giving rise to any visible branches to the corpora cardiaca. On entering the suboesophageal ganglion the two groups of five axons from left and right sides merge to form a single median group of 10 axons which are readily traced through the suboesophageal and thoracic ganglia. The abdomen of *Megoura* receives a meagre nerve supply in the form of a single abdominal nerve 12 μm or less in diameter which emerges from the posterior end of the fused thoracic ganglionic mass (abdonimal ganglia are absent). PAF positive axons were followed into this nerve, but on account of its small size individual axons could not be traced to their final destinations by light microscopy. No neurohaemal sites containing PAF positive material have been found in the abdomen. Indeed, it seems unlikely that the secretory product of Group I NSC would be transported intra-axonally from the brain to the abdomen, only to be released into the general circulation.

Hence it is probable that Group I NSC provide a directed delivery of their secretory product to abdominal effector tissues. These effector tissues may well be parts of the reproductive system itself, for fine branches have been observed from the abdominal nerve to the walls of the ovarioles. This arrangement offers a structural basis for a local neurosecretory control of morph determination in the ovarioles and may explain the apparent absence from the haemolymph of factors influencing morph determination.

The ability of as few as two Group I NSC to effectively regulate determination in all the ovarioles of operated insects may indicate that extensive branching of the neurosecretory axons occurs in the abdomen in order to provide all the ovarioles with a direct supply of secretory material. Alternatively, a more conservative explanation would be that the endings of the neurosecretory axons may consist not of specialized terminations on the effectors but of 'leaky' release sites in their proximity from which released material may diffuse over a wider area.

In summary, the secretory product of the NSC in the brain is not released into the blood from known neurohaemal organs such as the corpora cardiaca but is transported intra-axonally through the central nervous system to the abdomen where it may be delivered directly to the reproductive system. This arrangement enables the NSC to determine events in the reproductive system by local neurosecretory control rather than by release of a vascular hormone. Group I NSC are thus regarded as neurosecretory effector neurons of the photoperiodic morph determination mechanism.

NEURОSECRETORY DETERMINATION OF MORPH TYPE

No functions other than the photoperiodic control of polymorphism appeared to be impaired in aphids in which Group I NSC had been destroyed. This is consistent with the cytological homogeneity of the constituent cells of this group seen in sections stained with PAF (Steel, 1976a). These observations, taken together with the above discovery that any two NSC appeared able to compensate for the absence of the others, suggest that Group I NSC is functionally homogeneous.

It is not apparent from the above experiments whether the effect of the NSC is on the determination of virginoparae, oviparae or both. In order to investigate this point, the same lesions to Group I NSC were again performed, but this time on aphids which were not exposed to a subsequent change in daylength. Virginoparae which had experienced only LD were operated on immediately after the adult ecydsis as before, but returned again to LD. These insects proceeded to give birth to daughter virginoparae (in response to the preoperative LD regime) until the point in their progeny sequence at which they would have showed a switch in morph production had they experienced a change in daylength at the time of surgery. At this point, they switched spontaneously to production of oviparae even though they remained in LD. The ability to produce virginoparae had been destroyed in these aphids, which continued to make only oviparae for the rest of their lives. Clearly, intact Group I NSC are required for the determination of virginoparae under LD conditions.

In the converse experiment, virginoparae which had been reared in SD were subjected to the same operation at the same stage of development and replaced in SD. These animals gave birth to oviparae throughout their lives and showed no tendency towards spontaneous switching of morph production. Hence the NSC are unnecessary for the determination of oviparae in SD.

Thus, the NSC are required to elicit or sustain the determination of virginoparae in LD, but not of oviparae in SD. The action of these cells is therefore considered to be virginopara-promoting. It seems reasonable to attribute this promotive influence of the NSC to the release of their secretory product under LD but not SD. Consequently, the name 'virginoparin' seems appropriate for this new neurosecretory product of the insect brain.

It should be stressed that the rôle of this material is in the maternal control of developmental switching. The development of progeny once determined as virginoparae or oviparae by the mother is doubtless under intrinsic endocrine control and it is in these later stages of development that endogenous hormones, such as juvenile hormone, are doubtless involved. Indeed, it is possible that one of the actions of 'virginoparin' on the developing embryo consists of programming the daughter's corpus allatum for a particular pattern of secretory activity later in ontogeny. As has been observed by Lees (1966) the ovipara is a more extreme neotenic than the apterous virginopara. One of the actions of 'virginoparin' may therefore be to suppress the secretion of juvenile hormone from the corpora allata of the progeny.

PHOTOPERIODIC REGULATION OF NEUROSECRETORY CELLS

Group I NSC have been identified above as the effector of the photoperiodic morph determination mechanism. What can be said of the photoreceptors and photoperiodic clock which regulate them?

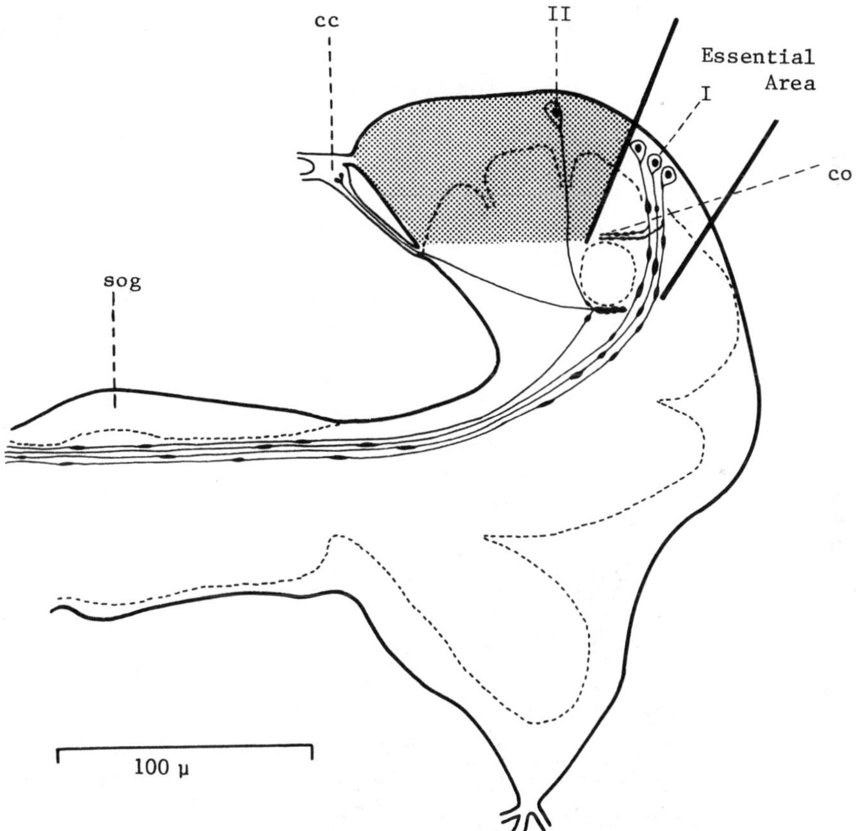

Fig. 3. Near sagittal section of the brain of *Megoura* to show the wiring of the neurosecretory cells of Groups I and II. The region between the two heavy lines is essential for operation of the morph determination mechanism. Stippled area has been experimentally destroyed without impairing the mechanism. I, II, cell bodies of Group I and II neurosecretory cells; cc, corpus cardiacum; co, collateral of Group I; sog, suboesophageal ganglion.

If all three components of the mechanism were located within the NSC, the mechanism would be expected to function independently of other neurons in the brain. It has already been seen that the mechanism is independent of very large areas of the protocerebrum, as summarised in Fig. 3. However, the presence of collateral branches of Group I NSC which terminate in the neuropile suggests that the cells are not autonomous, for it is through such processes that presynaptic input is received (Adiyodi and Bern, 1968; Maddrell, 1974).

The ability of the mechanism to function in the absence of presumed input to the NSC through the collaterals was examined in the following experiment. Parent virginoparae reared in SD to the adult ecdysis were subjected to deep lesions to the neuropile in the area around the endings of the collaterals. This was achieved by inserting the cautery electrode deeply into this region of neuropile through area C of the protocerebrum, which has been shown above to be dispensable. The animals were then transferred to LD to determine if they were able to respond to the post-operative change in daylength as in previous switching experiments. None of these operated aphids switched morph production. Careful histological autopsies revealed that the lesion around the collaterals did not reach the perikarya or axons of Group I NSC and the perikarya appeared cytologically normal. However, the normally abundant neurosecretory material in the axons deep in intact neuropile was depleted or lacking. Since this depletion occurred throughout the passage of the axons through ganglia remote from the lesion and in cells whose perikarya appeared intact, the depletion of stainable secretory material is unlikely to be a result of damage to the perikarya or axons of the NSC. Rather, it seems that the integrity of the collaterals is essential to maintenance of the normal secretory behaviour of the cells. This is consistent with the view that the collaterals represent the region which receives presynaptic input essential for the normal response of the cells to a change in daylength.

Efforts were therefore made to locate areas of the brain which might provide the NSC with this essential input.

Attention was first turned to the role of the existing optic machinery, which in *Megoura* consists of compound eyes, triommatidial eyes (Pflugfelder, 1936) and optic lobes (Fig. 1). Ocelli are absent in apterous virginoparae. Various surgical procedures were devised to produce visually blind aphids, including insertion of coarse cautery electrodes into the centre of the compound eyes, which produced areas of lesioning shown in Fig. 2 (area A), ablation of the distal optic chiasma between the compound eyes and optic lobes, and by mechanical disconnection of the eyes from the optic lobes. None of these aphids showed behavioural responses to light. However, none of these procedures prevented a normal switch in morph production in response to a change in daylength. Clearly, the eyes are not the photoreceptors of the photoperiodic mechanism in *Megoura*.

A variety of lesions were then made in the optic lobes by techniques similar to those used to create lesions elsewhere in the protocerebrum, except that in many cases the electrode was plunged deep into the neuropile, in order to produce almost complete coagulation of the optic lobe tissues (Fig. 2, area B). Of 30 animals so operated, only one failed to switch morph production in response to a change in daylength. Histological autopsy of this animal revealed a huge lesion on one side which extended into the medial protocerebral neuropile containing the axons of Group I NSC. This single response failure is therefore attributed to damage to the neurosecretory axons. The optic lobes themselves appear to be unnecessary for the operation of the photoperiodic mechanism.

The results of these experiments provide experimental confirmation of the conclusion reached by Lees (1964) from studies using localised illuminators that the eyes and associated optic machinery do not constitute the photoperiodic receptors in *Megoura*.

Having established that complete destruction of areas A, B and F and destruction of the cell bodies of area C (Fig. 2) are all inconsequential to the operation of the photoperiodic morph determination mechanism, attention was redirected towards the role of the anterior protocerebrum.

Ablation of perikarya in the lateral parts of the anterior protocerebrum (Fig. 2, area E) produced consistent failures of the mechanism. This area is close to the Group I NSC and in most animals a few NSC were damaged or destroyed. However, the extent of damage to the NSC was the same as, or less than, that sustained by the animals in which some of the NSC were destroyed without impairing the mechanism in earlier experiments. As it has been shown earlier that the mechanism is unimpaired by ablation of up to eight of the 10 NSC and only a few NSC were damaged in these experiments, it is difficult to attribute the results of lateral anterior lesions to damage of the NSC. It will also be recalled that the neurons among and between the NSC are also dispensable. It is therefore concluded that the effect of these lateral lesions is due to the destruction of lateral protocerebral neurons and not to a loss of NSC. The regions of the brain which have been found to be essential and inessential for the operation of the morph determination mechanism are summarised in Fig. 4.

Histological autopsies of the animals with lateral anterior lesions revealed the surviving NSC to be cytologically normal, but their axons were depleted of stainable secretory material throughout their length. This is precisely the same effect as was produced by damage to the Group I NSC collaterals. It therefore seems probable that the essential input to the NSC collaterals derives from neurons in the lateral protocerebrum. These results indicate that the NSC are not autonomous effectors, but are subject to presynaptic control by adjacent neurons. Since it is the release of the effector substance from the neurosecretory axons which appears to be controlled by these neurons, it is tempting to speculate that the input with which they provide the NSC derives from a neuronal photoperiodic clock located lateral to the NSC. This arrangement is summarised in Fig. 4. These neurons presumably also function as the photoperiodic receptors. However, no special cytological features have been observed in these cells.

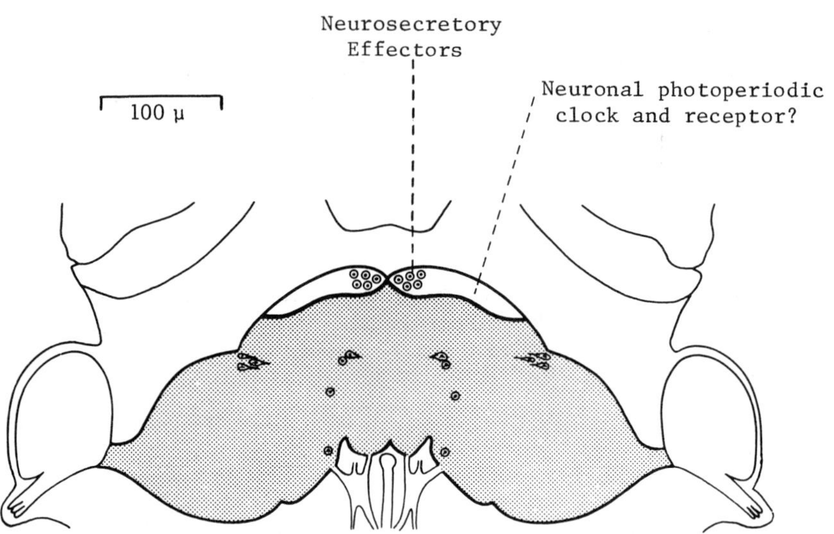

Fig. 4. Proposed localisation of the components of the photoperiodic morph determination mechanism in the brain *Megoura*. Stippled areas have been experimentally destroyed without impairing the mechanism. Unstippled areas are essential.

This proposed system has certain features in common with the only other insect system in which the components of a photoperiodic system have been examined. Williams (1969) found that the photoperiodic mechanism controlling diapause in *Antheraea pernyi* comprised the 'medial' NSC of the brain plus a more lateral area. However, Williams was unable to determine whether the essential feature of this lateral area was the lateral group of NSC or other adjacent neurons. In *Megoura*, optic lobe lesions frequently destroyed the Group V NSC, and these are probably equivalent to the lateral NSC in *Antheraea*.

In *Antheraea*, the effector regulates development of the organism which produces it, whereas in *Megoura* the effector acts only on the next generation. This may be related to the fact that the effector is endocrine in *Antheraea* (being the prothoracotropic 'brain' hormone) but probably locally delivered in *Megoura*. For, if as has been proposed above, one of the functions of the maternal factor in *Megoura* is to modify the programming of the endocrine system of the progeny, natural selection may have favoured a mechanism which prevents interaction of the maternal factor with the parent's own endocrine system.

REFERENCES

Adiyodi, K.G. and Bern, H.A. (1968). Neuronal appearance of neurosecretory cells in the pars intercerebralis of *Periplaneta americana* (L.). Gen. comp. Endocrinol. 11, 88-91.

Johnson, B. (1963). A histological study of neurosecretion in aphids. J. Insect Physiol. 9, 727-739.

Lees, A.D. (1959). The rôle of photoperiod and temperature in the determination of parthenogenetic and sexual forms in the aphid *Megoura viciae* Buckton - I. The influence of these factors on apterous virginoparae and their progeny. J. Insect Physiol. 3, 92-117.

Lees, A.D. (1963). The rôle of Photoperiod and temperature in the determination of parthenogenetic and sexual forms in the aphid *Megoura viciae* Buckton - III. Further properties of the maternal switching mechanism in apterous aphids. J. Insect Physiol. 9, 153-164.

Lees, A.D. (1964). The location of the photoperiodic receptors in the aphid *Megoura viciae* Buckton. J. exp. Biol. 41, 119-133.

Lees, A.D. (1966). The control of polymorphism in aphids. Adv. Insect Physiol. 3, 207-277.

Lees, A.D. (1967). The production of the apterous and alate forms in the aphid *Megoura viciae* Buckton, with special reference to the role of crowding. J. Insect Physiol. 13, 289-318.

Lees, A.D. (1972). The role of circadian rhythmicity in photoperiodic induction in animals. Proc. int. Symp. Circadian Rhythmicity (Wageningen, 1971), Pudoc, pp 87-110.

Lees, A.D. (1973). Photoperiodic time measurement in the aphid *Megoura viciae*. J. Insect Physiol. 19, 2279-2316.

Maddrell, S.H.P. (1974). Neurosecretion. In: Insect Neurobiology (ed. J.E. Treherne). North Holland, Amsterdam, pp 307-357.

Pflugfelder, O. (1936). Vergleichend-anatomische, experimentelle und embryologische Untersuchungen über das Nervensystem und die Sinnesorgane der Rhynchoten. Zoologica, 93, 1-120.

Steel, C.G.H. (1976a). The neurosecretory system in the aphid *Megoura viciae*, with reference to unusual features associated with long distance transport of neurosecretion. (In preparation).

Steel, C.G.H. (1976b). Some functions of identified neurosecretory cells in the brain of the aphid *Megoura viciae*. (In preparation).

Steel, C.G.H. and Lees, A.D. (1976). The role of neurosecretion in the photoperiodic control of polymorphism in the aphid *Megoura viciae*. (In preparation).

Williams, C.M. (1969). Photoperiodism and the endocrine aspects of insect diapause. Symp. Soc. exp. Biol. 23, 285-300.